The Bear Made Me Buy It
Product Advertising Bears

by Joyce Gerardi Rinehart

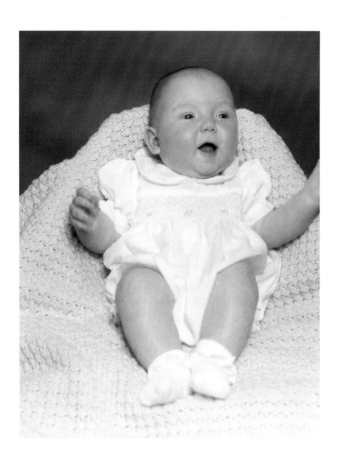

Dedication

This book is dedicated,
with eternal love and
"busting my buttons" pride,
to my first grandchild, granddaughter
Ashton Leigh Gerardi
born St. Patrick's Day, 1998,
and her parents,
Steven and Kelly Lee Catto Gerardi

Book Design by Anne Davidsen
Type set in Impress BT/Humanist 521 BT

ISBN: 0-7643-0734-7
Printed in China
1 2 3 4

Published by Schiffer Publishing Ltd.
4880 Lower Valley Road
Atglen, PA 19310
Phone: (610) 593-1777; Fax: (610) 593-2002
E-mail: Schifferbk@aol.com
**Please visit our web site catalog at
www.schifferbooks.com**

This book may be purchased from the publisher.
Include $3.95 for shipping.
Please try your bookstore first.
We are interested in hearing from authors
with book ideas on related subjects.
You may write for a free catalog.

In Europe, Schiffer books are distributed by
Bushwood Books
6 Marksbury Rd.
Kew Gardens
Surrey TW9 4JF England
Phone: 44 (0)181 392-8585; Fax: 44 (0)181 392-9876
E-mail: Bushwd@aol.com

Contents

Note

Preface

Welcome to the wonderful world of advertising bear collecting!

I can credit (or blame) an astute magazine editor for this book becoming a reality. In 1994, I wrote an article for her on Smokey Bear's 50th birthday, and another that I titled "A is for Advertising, B is for Bear" and she renamed "Huckster Bears." For that article she asked that I list books where her readers could go for more information. There weren't any books devoted to the subject. And searches through volumes on advertising items, teddy bears, dolls, toys, flea market finds, and antiques uncovered very few mentions of advertising bears. Molly's solution? "So, write one."

The idea took me by surprise and then took root. After four years of research, intensive "hunting for bear," and contacting bear collectors from Maine to California, the book blossomed into a much larger and more comprehensive volume (and joy) than I had imagined.

I hope that *The Bear Made Me Buy It* becomes a companion in your future bear-hunting expeditions and that you enjoy it as much as I enjoyed shopping for and writing it. - JGR

Acknowledgments

A book is never a one-person endeavor, this one has involved numerous individuals and companies. I hope the following list includes all who helped me, for their contributions are greatly appreciated. These people have added to the book's depth, beauty, and content while infusing me with enthusiasm while I wrote.

Grateful thank-yous go to:

MOLLY BLAYNEY, a super editor, who recognized the need and suggested the book,

PETER SCHIFFER for publishing me again,

NANCY SCHIFFER, a nurturing and knowledgeable editor,

GINNY KREITLER, member of UFDC and the Greater St. Louis Doll Club, for contributing a whole slide program of advertising bears,

The late JEAN LAUGHERY for sharing photos of three dozen of her advertising bear babies, and her zest for life,

NANCY VANSELOW for dozens of bears, encouragement, a museum jaunt, and proofreading,

DIANNE RINEHART HOUSER, CAROL and EDITH RINEHART FORD, and LAVONNE MORRELL for finding "beary" good bears for my collection,

and son GREG GERARDI, who planted the bear collecting seed in our family three decades ago.

Beary best wishes, multitudes of thanks, and bear hugs to these sharing, caring bear collectors who kindly sent me photos, allowed me to photograph their charming advertising bears, or supplied needed information:

Karen Albritton, Rue and Jean Berryman of Rujean's Collectibles Past and Present, Oneida Callaway, Judy and Stewart Callaway, Tiffany Chorney of F.A.O Schwarz, author Kerra Davis, author Ann Christensen Denney, photographer Mary Ann Callaway Dennis, Dee Domroe of the Doll Room, Nancyann Eckhart, Helen B. Evans of Billie's Emporium, photographer Vivian L. Gery, Nancy Harmon, bear columnist and author Dee Hockenberry of Bears N Things, Muriel Hoffman, Mary Duryea Kyviakidis, Nancy Kraus of Yogi Bear's Jellystone Park Campground, Gloria McAdams, Alison Hubbard-Miller and photographer Bill Miller, Peggy Monahan, photographer Tami Patzer, Shirley Pike of 2nd Time Around, Robert Reed of Antique & Collectible News Service, Doris B. Smith of 2nd Chance Home Furnishings, Shirley Taylor, James Thomas, Tony Tyndall, Bonnie and Larry Vaughan of Raggedy's & Teddy's, and Monica Wilson.

Thanks to Molly Higgins at Schiffer Publishing for "holding my hand" by phone during the final stages of the manuscript. And kudos to everyone at Ritz Camera for giving me such great service processing my zillion rolls of film.

Muchas gracias to these corporations, agencies, and individuals for their pleasant and helpful assistance:

Cooperstown Bears, Diane of Douglas Cuddle Toys, Grace Jackson of McPherson General Store in Cameron, North Carolina, F.A.O. Schwarz for "Truffles," Georgette Thomas of The Hugging Bear Inn, Leisure Systems Inc. on behalf of Hanna-Barbera® for use of Yogi Bear's image, Lorene Shiraiwa of North American Bear Company, Stephen Diehl and Robert Dagley of The Orvis® Company, George B. Black Jr. of Teddy Bear Museum of Naples, and special thanks to Craig Wolfe, president of The Blessed Companion Bear Company and Name That Toon.

And last, but definitely not least, grateful thanks to my strong support system: my sons who always cheer me on, my husband Al who feeds my morale and tummy during the book-birthing process, my co-workers at the paper and the mall, and members of the Port Charlotte (Florida) Doll Club and the Tuesday Writing Group.

Introduction

Ever since President Teddy Roosevelt DID NOT shoot a bear cub in Mississippi in 1902, Americans — and the rest of the world — have taken Teddy bears into their hearts and homes.

Advertising bears (also called logo, mascot, emblem, huckster, or spokes bears) are nearly "as old as the hills." Eons ago, cave-dweller hunters sketched bears on cave walls "advertising" their superior hunting skills. Centuries ago, before the populace could read, pictorial signs were necessary. British pub signs, swinging wooden plaques depicting "The Bear and Boar" and other establishments' namesakes, were in general use. No doubt an early American tavern or two had a bear in its name, and a sign which bore the bear's likeness. Two current day hostelries associated with bears are The Hugging Bear Inn in Chester, Vermont and Bear Flag Inn in Sacramento, California.

Extant examples of century-old American advertising items include the 1890 ad booklet "Bright Eyes and the Three Bears" given away by H.O. Oatmeal (whose advertising slogan was "I want some more.") In 1907 (five years after the famous Teddy and bear cub cartoon), Cracker Jack® produced a set of 16 cards called "Cracker Jack® Bears meet President Teddy Roosevelt." Cracker Jack® is still advertising with bears.

Although Smokey Bear is considered the senior statesman of advertising bears, he was a latecomer. Smokey posters appeared initially in 1945 and the first stuffed Smokey bear was issued by Ideal in 1952. One of his predecessors was from Kellogg's cereal, who, in 1925-26, had produced as premiums two sets of fabric cut-and-stuff figures of Goldilocks and the Three Bears.

Advertising dolls, bears, and other premiums originally were obtainable primarily through the mail in exchange for box tops or labels and mere pocket change. Some small items were enclosed in specially marked boxes of cereal or washing powder. Since they were considered temporary and disposable, high quality was not a concern. A very few advertising critters, such as the ShopRite grocery store bear and baseball team bears, were available for on-site purchase. Others, like the Tootsie Roll® bear, have been offered in gift catalogs, usually for a brief period, priced from $10 to $30.

In the past five years, advertising bears unobtrusively climbed to the top of the social ladder and are appearing in the "best" collections. What has made them more socially acceptable? Manufacturers such as Douglas Cuddle Toys now produce a series of classic, limited edition, high quality bears touting the products of long time, high profile advertisers such as Hershey™s (chocolate), Campbell's® (soup),

Lipton® (tea), Morton® (salt), the Saturday Evening Post (magazine), Schwinn (bicycles), Pepperidge Farm (bread, cookies), Ben & Jerry's (ice cream), and Oreo (cookies). In 1984 North American Bear issued a blue bear with an Aunt Jemima® apron. They also produced bears for Cracker Jack®, Saks Fifth Avenue, and Heinz®. In addition to bears promoting sports teams and Olympic games, bears promoting Colonel Sanders (chicken), Cracker Jack® (caramel corn), Harley Davidson® (motorcycle), and Chiquita® (banana) appear in Cooperstown Bears current catalog.

Whatever your collecting budget, you'll find cuddly advertising bears in your interest area and price range.

Selection

Rule Number One: buy only those bears you love and want to share your home space with (unless you are buying for resale or just can't resist the price, thus obtaining bears inexpensively for trading purposes). While gathering information for this book, I bought nearly every advertising bear I saw. Some will remain in residence. Others will be adopted out to good homes.

Selection criteria differs for each collector, whether for a general collection or one or more of the following categories:
o High quality limited editions
o Specific maker
o Specific artist/designer
o Product bears
o One specific product
o Tourist bears, domestic or foreign
o Sports bears or
o One specific team
o Hobbies: golf, tennis, skating, gardening
o Specific types: Panda, koala, colored, dressed, undressed
o Specific characters: Yogi, Smokey, Paddington, Pooh, Berenstain, Snuggle
o Celebrity bears
o Storybook bears
o Corporate tags
o Message bears (Happy Birthday, I Love You, etc.)
o Bears that demand to be a part of your family

Identification

I recall a fashion article of the 1950s suggesting that we drape our coat, jacket or sweater over our chair at a restaurant in order to show off its label. The writer bemoaned it was the only way to let the expensive maker label be seen. Forty years later it's nearly impossible to find a pair of jeans or sneakers, a shirt, T-shirt or hat that does NOT proclaim the product or designer's name in read-across-the-room sized letters.

When hunting for collectible advertising bears, remember that not all bears will have logos emblazoned conspicuously on their chest. Some, like the Elizabeth Arden bear, have only discreet sewn-in seam tags.

Some bears are marked in several places, others in only one place. You'll find advertising information on:
o Paper tags
o Body tags
o Ear buttons
o Chest buttons
o Plastic tags
o T-shirts
o Neck ribbons
o Body sashes
o Hats (often baseball or knitted caps)
o Clothes (jeans, sweater, overalls, dress, shirt, pocket, belt)
o Name embroidered on a foot
o Name on a patch on the backside
o A product (Coke bottle, Hershey's bar) the bear is holding
o A team pennant the bear is holding
o Mark on neck if vinyl head.

If you find a bear you think is a duplicate, check its tag. It may have been made by a different company, have a different date, or have been produced in another country. Some tags, like those of Gund and R. Dakin, will conveniently give copyright dates; others are undated and require research on your part. For example, undated Gund bears can be sorted out according to the address shown on the tag: from 1940 to 1956 the address was New York City; from 1956 to 1973 it was Suydam Street, Brooklyn; from 1973 to 1987 their location was Ethel Road in Edison, New Jersey; the address since 1987 has been Runyons Lane, Edison.

Price Determinations

In addition to condition of the bear and popularity of the product, several other factors affect the price/value of an advertising bear.
Prices will be HIGHER if:
o It is MIB (mint in box), MIP (mint in package), NRFB (never removed from box)
o The company is no longer in business
o The product is no longer available
o All tags: body, paper, plastic, are intact
o All clothes and accessories are present
o It is accompanied by the original ad or premium offer
o In sports, if the uniforms or hometowns have changed or their names or mascots are no longer politically "correct" (i.e. Braves, Chiefs, Indians, Redskins)
o It is a limited edition
o It is made by a top-of-the-line company
o It is of superior quality
o It is an older bear (1890s through 1960s)
o It was available for a very short time
o It is part of a package with dated ads picturing the bear.
o The bear has a particularly attractive or appealing outfit or appearance
The price generally will be LESS if:
o Identifying T-shirt or tags are missing
o The only identification is a paper tag
o The only identification is an undated body tag
o The product or service is not well known
o Interest is strictly regional
o Department store bears tend to be less pricey on the secondary market than product bears

Of course, bear and advertising collectors have, for years, been picking up pre-owned advertising bears at garage sales, thrift shops, and flea markets for extremely reasonable prices: 25 cents to $5. Antique shops tagged them at $5 to $15. But with the advent of "bear fever," this has escalated to $15 to $35. The diversity of advertising bears available is encouraging and challenging to the collector.

Purchase Sources

If you'd like to buy some advertising bears, or if you just want to dream and drool over them, send for catalogs from these companies:

o American Originals, PO Box 85098, Richmond, VA 23285-5098 (brand name gear)

o L.L. Bean Inc., Casco St., Freeport, ME 04033

o bear-in-mind inc., 53 Bradford St, W. Concord, MA 01742 (since 1978)

o The Blessed Companion Bear Co, 28 Mountain View Ave, San Rafael, CA 94901

o Coca-Cola™ Catalog, 2515 E. 43rd St, PO Box 182264, Chattanooga, TN 37422

o Cooperstown Bears, 1275 Busch Parkway, Buffalo Grove, IL 60089

o Douglas Cuddle Toys, Box D, Keene, NH 03431-0716

o Lands' End Inc., One Lands' End Road, Dodgeville, WI 53595-0001

o The Lighter Side™ Co, PO Box 25600, Bradenton, FL 34206-5600 (since 1914)

o North American Bear Co, 401 N. Wabash, Suite 500, Chicago, IL 60611

o Woodland Enterprises, 310 N. Main St, Moscow, ID 83843 (Smokey Bear items)

o Carol Wright Gifts, PO Box 8502, Lincoln, NE 68544 (Team NFL, etc.)

Bear collector magazine ads, newspaper ad inserts, and grocery store packaging are still viable sources of advertising bears and other advertising items. Be on the lookout for these send-away sources.

Addresses of Interest

o Good Bears of the World, PO Box 13097, Toledo, OH 43613

o Hot Foot Teddy Newsletter, PO Box 1416, Westwood, CA 96137 (Qtrly, $20/yr., Smokey Bear information)

o The Hugging Bear Inn & Shoppe, Main St, Chester, VT 05143, 802/875-2412 (1850s Victorian mansion with tower and carriage house. A bear in every bed for guests to hug. Barn shop sells a diversity of bears. Inn hosts Vermont Teddybear Artists day in September each year.)

o Richard Yokley (Smokey Bear collector/researcher), Box 718, Bonita, CA 91908-0718 (Will answer questions or share information)

o Teddy Bear Guide Book (annual publication), Barbie Hampton, editor, 15172 Koyle Cemetery Rd., Winslow, AR 72959

o Teddy Bear and friends (bi-monthly magazine), Cowles Enthusiast Media, 6405 Flank Dr., Harrisburg, PA 17112

o Teddy Bear Review (bi-monthly magazine), Collector Communications Corp., 170 Fifth Ave., New York, NY 10010

o Teddy Bear Museum, 2511 Pine Ridge Rd., Naples, FL 33942 (Museum Director George B. Black, Jr. is an expert witness in court cases involving advertising bears.)

The Advertising Bears

1. Company Bears A to Z

Bears advertising products, services, companies, organizations, and programs are arranged in alphabetical order according to company name. Although koalas are marsupials and pandas are members of the raccoon family, they are included since most of us consider them "bears." You'll find named bears listed in the book's index.

A

A&W ROOT BEER - 10" A&W "Great Root Bear" hand puppet from 1976 has only his A&W chest patch and a "Made in Taiwan" tag. It was an in-house promotion advertised mostly by posters in the restaurants. His original price was $1.00. The A&W Root Beer restaurants celebrated their 75th anniversary in business in 1994. Hand puppet current value $18-$25.

A&W ROOT BEER - 12" "Great Root Bear" on left is untagged, but appears to have the 1975 face. The 14""Rootie" on the right, from the same time period, has lost his shirt patch. (Some glue still remains.) His leg tag says: "Made in Korea, For Brasure Toys, Monte Bello, Calif. 90640." Contents and "surface washable" are on the reverse. Both bears are plush with felt features, felt footpads, long tails, and clothes that are part of their bodies. In 1975 their price at root beer stands was $3.75. By 1985 they were "widely available in Canada and the United States" for about $8.99. *Information courtesy of Jacqueline Wilson.* Current values $25 for 12", $18 for 14" without patch.

A&W ROOT BEER - The 4" bear on the left is made from different sizes of plush pompoms, and has a gold cord on his head for hanging. Gold metallic paper tag on cord: "A&W Great Root Bear ®, A&W Restaurants Inc., Made in Korea." The 7" seated plush bear in the center is tagged "Canasia Toys & Gifts Inc., Downsview, Ontario M3J 2C4, Product of Korea." He has only a nub of a tail. The 6" bean bag "Root Bear" on the right has a brown velour body and flannel-like non-removable shirt and hat. A&W patch is plastic. His leg tag reads: "Alpha Kids, A Division of Alpha Corporation, Toronto, Canada." Tag reverse: "© 1997 A&W Restaurants Inc., Manufactured in China by Quairico Industrial, Ltd. Hong Kong." The original price at the Root Beer stand in the Mall was $4.99. Current values: 4" $6.50-$10, 7" $8-$12, 6" bean bag $6-$12.

A&W ROOT BEER - Even with one ear missing, this 13" seated "Great Root Bear" could not be left behind. His felt mouth is nearly "loved" off. Eye brows are black plush, nose black flannel, and eyes plastic. His plush clothes are part of his body. Tush tag: "Canasia Toys & Gifts, Inc., Downsview, Ontario M3J 2C4, Synthetic fibers, Product of Korea." Note this Rootie has a different look from the earlier bears. As is $15.

ABEL CREATIONS COMPANY - Circa 1990, 6" seated white bear, dark brown velour nose, plastic eyes, white neck ribbon, royal blue knit sweater, knitted in white on front "Hampton" and "Pirates" on back. Tush tag: "Acme, Made in Korea." Paper tag: "Educated Bear" and bear drawing ™ on front. Tag reverse: "Hi! I'm so well educated that I can write any letters on your sweater. Order me from your favorite store. Abel Creations Company, 123 N. Hamilton Road, Columbus, Ohio 43213." *Courtesy of Shirley Taylor.*

AGWAY (seed and grain merchants co-op) - 16" standing dark brown bear with shaved muzzle, wearing tan barn boots, blue bib overalls, red plaid flannel shirt. Red hat says: "Agway." Tag: "Patriot Bear" by J. J. Wind. This was an in-store offering Christmas 1994, original price $25. *Courtesy of Jean Laughery, photo by Vivian L. Gery.*

ADMIRAL (refrigerator) - Circa 1950s, 13" seated white bear, red open mouth, blue wide wale corduroy shirt is part of his body. Gold script embroidery on the shirt reads: "Admiral Dual Temp." It was a gift with the purchase of a refrigerator. *Courtesy of Ginny Kreitler.*

AMBASSADOR (Division of Hallmark Cards) - 9" seated light brown plush bear, with amber plastic eyes, black velour nose, and slightly darker brown back paw pads. He wears a red plaid taffeta neck ribbon. Paper ear tag reads: "Ambassador (heart), Please give lots of tender care To your lovable, huggable holiday bear!" Tag reverse: "Mfd for and © Hallmark Cards Inc, KC, MO, Made in China." Side tag repeats the information on the paper tag reverse. *Courtesy of Mary Duryea Kyviakidis.*

ALLIED STORES - 16" all white acrylic plush bear, felt nose, plastic eyes, wearing red knitted cowl neck sweater and long red knitted stocking cap. Leg tag: "Sold by Allied Stores International Inc.,// NYC, NY 10036// Made in Korea" Undated, $4-$8.

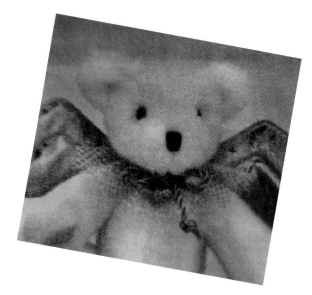

AMERICAN GREETINGS - White plush bear with black plastic eyes and nose, gold fabric angel wings and paw pads, golden cord neck bow. December, 1997, Eckerd ad: "Save $4, Angel Bear $2.99 with purchase of 3 American Greetings Cards. $6.99 value. Limit 3 offers per customer."

AMERICAN EXPRESS (credit card) - 8" seated butterscotch plush bear, beige snout, ear linings, and back paw pads, plastic eyes and black plastic nose, black floss mouth. Has chubby tummy and rounded humped back which makes head protrude forward. Wears turquoise T-shirt which says in white capital letters: "American Express, Don't Leave Home Without Me." Missing leg tag. Date unknown. $12-$20

ANHEUSER BUSCH (beer/snacks) - Beige plush "Eagle Snack Bear," 1986, made by Gund, body tag reads "bialosky 1982-1984." Long red satin neck ribbon reads: "Honey Roast Eagle Snacks," (honey roasted nuts). The company began in 1860. *Courtesy of Ginny Kreitler.*

ANHEUSER BUSCH (beer) - 6" standing (measured from top of hat), fully jointed, golden brown bear, black plastic top hat, pink felt ear linings, plastic eyes and nose, red shirt with "A" stylized logo, black bow tie glued on, gold hanging cord on top of hat with round gold metallic tag: "Honey Bears" and bear logo, Tag reverse: "©1988 Kurt S. Adler, Handcrafted in Korea." (Kurt Adler is noted for his wooden Raggedy Ann and Andy Christmas ornaments.) $5-$8

ANTIQUE & COLLECTIBLE NEWS SERVICE - "Antique Andy," Antique & Collectible News Service advertising bear. Dark brown with lighter paw pads. Shirt says: "Antique & Collectible News Service, Acns@aol.com, 765-345-7479." The Service, located in Knightstown, Indiana, supplies articles and photographs to magazines and newspapers. *Photo courtesy of Robert Reed.*

ANTIQUE AUTO COLLECTORS ASSN, Eastern Division - 12" gray-brown plush bear, green and white T-shirt features a 1934 Ford, red plaid neck bow. Tag: "Little Cuddlers ™, Douglas Co. Inc. 1988" Promotion: Hershey, PA Region Bear at National Fall Meet, October 6-8 in 1988. Original price $24. *Courtesy of Jean Laughery. Photo by Vivian L. Gery.*

ARAMIS (men's after-shave) - 7" high bear's red shirt says Aramis in small letters, feels like a bean bag In 1981 bear came with a bottle of after-shave and was called "the executive scapegoat." *Courtesy of Ginny Kreitler.*

ELIZABETH ARDEN (cosmetics) - 8" seated unjointed bear of tan lambs-wool-like fur with velvet paw pads, amber plastic eyes, and black plastic nose, straight back and rounded tummy. Wears taffeta neck ribbon of green, red, and blue plaid. Back ribbon tag: "Made with love exclusively for Elizabeth Arden in China (PRC)." No date. Was probably a premium with purchase of cosmetic products. $10-$15. .

AT&T - 10" seated, all white plush bear, no tail, plastic eyes, teardrop shaped brown velour nose, Bear holds stuffed black satin telephone receiver. Red T-shirt says (in white capital letters) on front: "Make The Distance Bearable...Call!" In white capital letters on the shirt back it says: "Telebear." Tush tag: "AT&T," Tag reverse: "Made in Taiwan." Bell Telephone began in 1877. $12-$20.

AUTOMATED DATA PROCESSING (ADP) (payroll service) - On the left, 10" chocolate brown bear with pointed light yellow snout, ear linings, belly and back paw pads. His removable red and white shirt says "ADP - We Care About You." Foot tag: "B.J. toy." A gift to payroll customers in 1986. SHELTER INSURANCE CO (life, health, car insurance) - On the right, 9" seated chocolate brown bear, white shirt with blue printing has company logo "Shield or Shelter" and a P.S. for personal service. *Courtesy of Ginny Kreitler.*

AVON PRODUCTS (cosmetics, door to door) - 20" beige bear, wears red shirt with "Avon" in white, shirt is part of his body. Bear was presented to Avon team leaders Christmas 1979. *Courtesy of Ginny Kreitler.*

Left: AVON PRODUCTS - 16" butterscotch plush bear, small plastic eyes, brown velour nose, tan felt back paw pads, dressed in red and blue vertically striped non-removable night shirt and night cap, arms and body are of the same fabric. Drop seat conceals battery pack and Velcro® closure. Seat tag: "Created Exclusively for Avon in China," Tag reverse: "Avon Products, Inc. Distr. New York, NY 10019, © Avon, All Rights Reserved." Undated. Mechanism is not functioning, activity is unknown. As is $6-$12

AVON PRODUCTS - 14" standing bear, white plush head and hands (sewn together in prayer position). Long-sleeved aqua flannel sleeper is part of bear's body. Sleeper has pastel plaid collar, cuffs, drop seat with pink felt buttons, and nightcap with yellow felt flower-shaped tassel. Floss sleep eyes and mouth, black velour nose, pink rouged cheeks. Drop seat fastens with Velcro® to reveal battery box. Press on bear's paws and he recites "Now I lay me down to sleep ..." Leg tag: "Avon Prods Inc Distrib//NY, NY 10019//© Avon 1996 - All Rights Reserved// "Created for Avon in China." $10-$20.

AVON PRODUCTS - 19" brown plush bear, snout lighter brown, plastic eyes, velour nose, open mouth lined with pink, plush front paws. Green print (gold bears and white stars) footed sleeper with drop seat (holding battery pack) forms the bear's body. Sleeper has knitted red cuffs and neckline and red felt "buttons" on drop seat which fastens with Velcro® strips. Tag: "Avon Products Inc. Distrib.// New York, NY 10019//(c) Avon 1993..." Reverse: "Created Exclusively for Avon in China." Mechanism is not functioning, but mouth is built so that it might have talked. As is $6-$12

AWANA CUBBIE - On the right, 11" tall, gold plush bear, googly eyes, large black pompom nose, red felt tongue, wearing blue felt vest with "Awana Cubbie" on it. Company, product, and date unknown. On the left 8" seated brown bear with beige snout and back paw pads, pink satin shield on chest says "Transderm-Nitro" (nitroglycerine medication for heart problems). *Courtesy of Ginny Kreitler.*

BASKIN ROBBINS INC. (ice cream) - 6" raspberry plush. Bear holds plastic spoon with legend "You Make Me Melt!" Teddy sits in a Dixie ice cream cup marked "Baskin 31 Robbins." Tag: "Applause 1987, Item 12074, "Lovin' Spoonfuls." This was an in-store promotion in 1987. *Courtesy of Jean Laughery. Photo by Vivian L. Gery.*

BALLOONS FOR YOU - 7" seated light brown plush bear, small plastic eyes, tiny black plastic nose, red neck ribbon. Tush tag: "CH1351 BEAR, Balloons For You © 1997." Tag reverse: "Balloons For You, Lumberton, NC 28358," There is glue on the back of his neck where a balloon may have been fastened. $5-$8

18" L.L. Bear with his companion bear "Ellie." Several other character bear friends are also available. *Courtesy of Jean Laughery, Photo by Vivian L. Gery.*

L.L. BEAN (mail order catalog from Freeport, Maine store) - 18" chocolate brown soft acrylic plush bear, with an "L.L.Bear" patch on his red hooded parka. His backpack is for carrying "bear essentials." Can stand upright on his big feet. Foot tag: "Cuddle Toys Classic by Douglas Co., Keene, NH 03431" He was designed to embody the characteristics of "loyalty, steadfastness, devotion to the outdoors, sportsmanship, and dependability." *Courtesy of Jean Laughery. Photo by Vivian L. Gery.* Purchased by mail order in 1991 for $39.00.

Back view of 18" L.L. Bear with his backpack used for carrying "bear essentials." *Courtesy of Jean Laughery. Photo by Vivian L. Gery.*

L.L. BEAN - 18" chocolate brown bear wearing turquoise-trimmed purple shirt and a red waist pack with his very own book "L.L. Bear's Island Adventure," by Kate Rowlinson, illustrated by Dawn Peterson. Several other L.L. Bear storybooks are available. Bear made by Douglas Company, 1993.

BEAR BRAND HOSIERY COMPANY- Prior to the 1930s Bear Brand Hosiery offered a family of three brightly printed muslin bears: Mama, Papa, and Boy Bear. Each was 9" tall and "carried" a different (printed on) box of the company's hosiery. Mama is pictured here. *Courtesy of Nancy Harmon.* $45-$75.

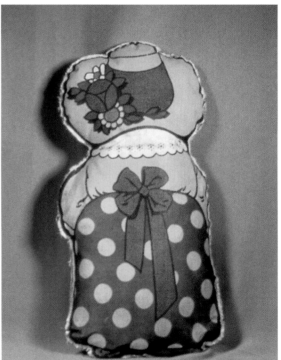

Above: Back view of Bear Brand Hosiery's Mama Bear, showing off her colorful hat. *Courtesy of Nancy Harmon*

Left: BELKS (department store) - 15" seated all white plush bear with nickel-sized plastic eyes, black velour nose, floss mouth. Dressed in a red knit sweater and mini skirt. Knitted into the sweater is "Belkie" in white and "88" in green, skirt is trimmed with white lace at hemline. Leg tag: "Outside Pile Outfits 100% Acrylic, Made in Korea." Boy bears also were available. *Courtesy of Muriel Hoffman.* $10-$15.

BENETTON STORES (fine line of sweaters and shirts) - 8" seated soft white bear, sold 1985-1987. His white shirt with blue trim is removable. Left hip tag: "Benetton." The bear came in several sizes. *Courtesy of Ginny Kreitler.*

BEN & JERRY'S - 17" Ben & Jerry's Ice Cream Bear wears a typical "hippie" outfit: "shades," tie- dyed Ben & Jerry's Ice Cream shirt, and dungarees with a peace symbol patch. Of course, he carries a carton of his favorite ice cream flavor "Ben & Jerry's Chocolate Chip Cookie Dough." Made by the Douglas Company, 1998, limited edition of 1,800. Issue price $105.50.

B.P.O.E. (Benevolent and Protective Order of Elks) (fraternal lodge) - 6" seated golden bear, tan snout, ear linings, and back paw pads, plastic eyes, velour nose, red hanging loop on head. Dressed in navy blue T-shirt picturing an elk head within a BPOE emblem on the shirt front. Bottom tag: (bear logo) "ASI 62960, Made in China." Courtesy of Carol and Edith Rinehart Ford. $8-$12.

BIG BEAR SUPERMARKETS (grocery chain) - Light brown bear with tail, on all fours, 6.5" high and 11" long. The black blanket on the bear' s back announces their "50th Anniversary, 1934-1984." and "Quality Supermarkets since 1934." This is the only bear the Columbus, Ohio company ever ordered production of. The chain was sold shortly after the 50th anniversary, but the name was retained. Big Bear's slogan still is "Give 'Em a Big Bear Hug." *Courtesy of Karen Albritton.* $20-$25.

BIOLAGE® - 13" standing, all white plush bear, amber plastic eyes, brown velour nose. Green neck ribbon says: "Biolage" in white. Leg tag: "Manufactured by Pam & Frank Industrial Co Ltd// Distributed by Matrix Essentials Inc.//Made in China." Paper ear tag front says: "Hello, My name is Bobbi Bear." Inside the folded tag it reads "When you take Bobbi home, please give him lots of his favorite things: hugs, wild flowers, stories, sunbeams, honey." On the paper tag back: "System Biolage® by Matrix® Essentials. $6-$12.

BOSCO - 14" seated brown bear, yellow T-shirt says "Bosco Bear" in brown; felt nose, black plastic eyes. Tag: "1979, Eden Valley, Minnesota, Animal Fair." *Courtesy of Dee Domroe.*

BOSCO (chocolate/malt flavoring for milk with iron added) - 13.5" tall bear wearing a white felt vest which says "Bosco" in red. Appears to be 1950s vintage. Has large felt eyes with pie slice pupil cutout, plastic snout with red mouth and nose freckles. Bosco's slogan from 1942 "Bosco and Milk, The Milky Way to Health." *Courtesy of Ginny Kreitler.*

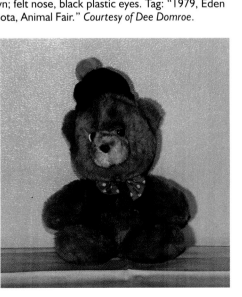

Left: BRACH'S CANDY - 11" tan plush "Ted E. Bear" "the bear who slept through Christmas" starred in national TV specials prior to Halloween/Thanksgiving/Christmas 1981. Wears blue felt vest and knitted cap with green pompom, green polka-dot bow tie. Tag: "Animal Fair, Eden Valley, MN 55324." Paper hang tag: "1981 The Lefave Company." The bear was a premium offer in 1981 for $10 plus one package label from Brach's Candy Mix. *Courtesy of Jean Laughery, Photo by Vivian L. Gery.*

BRAWNY PAPER TOWELS - On left 9.5" beige bear with lighter tan snout, ear linings, and paw pads. Yellow neck ribbon says "Brawny" in red. Leg tag: "Barker" and "© Russ, Berrie." On right 13.5" golden brown bear, removable red plaid shirt. Tag: "James River, 1987." *Courtesy of Ginny Kreitler.*

BRITISH STERLING (after shave/cologne) - Bears by Gorham, bought in 1986-1987 at a Sears store. A British Sterling purchase was required. The 15"standing, fully jointed tan plush Lady Sterling is elegantly dressed in a white blouse, red velvet jacket, black velvet skirt, and white petticoat. The black brimmed felt hat has a space for her ears. A ribbon on her foot says "Lady Sterling." The clothes are well made and removable. As are the clothes of 14.5" "Master Sterling" (marked on his right foot) who was issued in 1987. He, too, is fully jointed. He's dressed in a black jacket with red lapel trim, a white shirt, red "old school" tie, and black school boy cap with opening for ears. There is also a Lord Sterling (not shown). *Courtesy of Ginny Kreitler.*

BURDINES - 15" seated all white plush bear with plastic eyes and velour nose. Dressed in cotton swimsuit of a tropical print with pink flamingos, green alligators, blue dolphin, palm trees, shells, waves and banners "The Sunshine State." Paper tag "BEARDINES" (3 times) above and below a smiling white bear face in sunglasses with a red hibiscus by its ear. Tag reverse: "Hello, I come to you Bearing warm holiday greetings from a very special friend. I'm warm, cuddly, but I need big Bear Hugs to keep me happy.// Bear in mind, too, you may sponge me and air dry, brush me with a good stiff brush to keep me soft and cuddly." Leg tag: "Burdines exclusively" Tag reverse: "Francesca & Co. ™, Fairfield, Iowa, Made in Korea." Sunglasses replaced, missing ear flower. *Courtesy of Muriel Hoffman.* $15-$20.

BURDINES (department store) - 14" seated all white plush bear, large head, plastic eyes, black nose, black floss mouth, dressed in aqua sunsuit printed with palm trees, pink flamingo, and "Fla" in yellow. White satin ribbon tush tag says (in blue) "Burdines//Exclusively 1987", tag reverse reads: "Francesca & Co.™// Fairfield, Iowa// Made in Korea." A white thread near left ear may have held a paper tag or a flower. $15-$18. An identical Burdines bear was found with no clothes, no date on tag, and a short elastic loop on right ear. Bare bear's price $5-10.

C

Campbell's Soup

CAMPBELL'S ® SOUP COMPANY - Butterscotch-colored seated hump-back bear, 9" tall by 10" wide, wears a white chef's hat, red neckerchief, and a bib-style apron with vintage Campbell's Kids on it. (The Campbell Kids, with no names, were created in 1904 by artist Grace Drayton, well known to doll collectors.) Great attention was given to authentic detail. He holds a plastic cup of Campbell's tomato soup. His red molded plastic tongue seems to be "licking his chops." Paper ear tag is a replica of a red and white can of Campbell's tomato soup. "Made in China." "© 1997 Blessed Companion Bear Co.™" (One of the first three bear designs to debut this new bear company.) Limited edition of 5, 000. Issued March 1998 at $29.95 plus shipping.

CAMPBELL'S ® SOUP - 22" husky brown bear in removable T-shirt that announces his preference for "Campbell's Teddy Bear Soup." (See Bearly Edibles, for soup can packaging.) This chunky bruin has plastic eyes, leatherette nose and toes. *Courtesy of George B. Black, Jr., Teddy Bear Museum of Naples.*

CAMPBELL'S ® SOUP COMPANY - From second limited edition (1,200 pieces) 15" "Scottish Lass" Bear"® accompanied by "Bell" her snow-white Scots terrier. "Lass" is dressed in a traditional plaid kilt and matching tam. Logo on her jacket is a little Campbell's Soup can. Jabot and cuff trim of white eyelet. Suede paws, hand embroidered nose. Individually numbered and boxed. Made by Douglas Cuddle Toys for their Company Classics series, 1995.

CAMPBELL'S ® SOUP COMPANY - 15" Campbell's cheerleader, fully jointed, girl bear (3rd limited edition), wears red and white pleated skirt, red and white saddle shoes, a knit white wool sweater with "Campbell's" in red script, red ribbons on each ear, and red and white pompoms in each hand. Distressed curl plush, suede paws, hand embroidered nose. Individually numbered and boxed. Limited edition of 1,800, made by Douglas Cuddle Toys for their Company Classics series, 1998. Issue price $96.50.

CAMPBELL'S ® SOUP COMPANY - Two sizes (15" and 11") of Campbell's "Souper Chef" bears, wearing chef hat that says in red: "M'm! M'm! Good!" Aprons are replicas of red and white Campbell's Soup cans. Each wears a red and white plaid neckerchief. The 15" bear is fully jointed, has suede paws, hand-embroidered nose, distressed curl plush, and comes with a wooden spoon. A limited edition of 2,400, individually numbered and boxed, was offered in 1995. The 11" is fully jointed and his apron is silk-screened. Made by Douglas Cuddle Toys, Keene, NH, for their Company Classics series, 1995. In 1998 Douglas introduced a new 5" miniature line which included a tiny Campbell's Chef Bear. Issue price $16.50.

CARNATION COMPANY - 8.5" white plush bear wears red knitted hat, mittens, and scarf, sits in a ceramic mug that says "Carnation Hot Cocoa Mix." Tag: "Carnation Hot Cocoa Mix, Plush Promotions, Calif. USA." Promoted as "Cocoa Bear 'N' Mug Offer," in 1988, price $4.95 plus 2 UPCs. Offer expired November 30, 1989. *Courtesy of Jean Laughery, Photo by Vivian L. Gery.*

CELESTIAL SEASONINGS ® NATURAL HERB TEA - 18" brown bear sitting with crossed legs. Dressed in gray nightshirt and red nightcap. Shirt embroidered in red: "Celestial Seasonings Sleepy Bear." Embroidered closed eyes. Tag: "Trudy" and "Celestial Seasonings Herb Tea, Collector Series 'Sleepy Bear' #004370 of 5,000." Signed "Mo Siegel, President." Available in 1986 for $25. *Courtesy of Jean Laughery, photo by Vivian L. Gery.* Currently $50-$75.

CAR QUEST - 11" seated cinnamon brown plush bear with tan snout, ear linings, and back paw pads. Plastic eyes and black plastic nose with indented nostrils, floss mouth, rounded hunched back, chubby tummy, flat tail, seamed middle front and back. Wears red flannel pullover shirt. Outside shirt front lettering reads: "Car Quest//I (heart) You." Inside shirt front white lettering reads: "Alpha Phi." Tush tag: "Roly-Poly//Bear //California Stuffed Toys." Reverse: "Handcrafted by// California Stuffed Toys//Los Angeles, CA//...Made in Korea." Round paper tag: "California Stuffed Toys - made with love" and seated bear. Paper tag reverse: "My Name is Roly-Poly Bear, No. 3416, by © California Stuffed Toys, Los Angeles, CA 90023." (This company also made Winnie the Pooh bears.) Undated. $5-10

CHILDREN'S BOOK-OF-THE-MONTH CLUB - 8" seated tan plush bear with plastic eyes, and brown plush upside-down triangle nose, dressed in red knit sweater with round felt printed patch on the front. Patch shows children reading beneath a tree and the words: "Children's Book-of-Month Club." Seat tag: "Hachi International Inc., Made in China." Tag reverse has cleaning instructions. $5-$10.

CHILDREN'S PALACE (toy store) - 19.5" "Peter Panda" is their spokesbear. He was sold in their stores in 1986. Dressed in red and white striped shirt, blue overalls with yellow shoulder straps and a red patch on one knee. *Courtesy of Ginny Kreitler.*

CHRISTMAS CLUB (bank savings program)- 8" seated white bear, red flannel vest says in black: "Christmas Cub™." Vest has Velcro® closure in back. Olive green flannel bow tie, black plastic eyes and nose, red flannel Santa hat with white ball tassel and white rim. Tush tag: "Made exclusively for Christmas Club, A Corp.// R. Dakin & Co, 1982// San Francisco, CA// Product of Korea." The bear's bottom is weighted. *Courtesy of Shirley Taylor.*

CHIQUITA® BRANDS INC. - 15" Chiquita® Banana bear, limited edition of 1,000 by Cooperstown Bears as part of their Americana Collection. Listed in their 1998 catalog. Fruit bowl styled hat, yellow gown with red, yellow, and green ruffles at sleeves and hem. The Chiquita ® Banana trademark symbol was designed in 1944 and has been in continuous use since then.

COBBIE CUDDLERS ® (shoes) - On left 11" dark brown bear with beige snout cost $3 in 1984 with the purchase of a pair of Cobbie Cuddlers shoes. Tag on left rear: "Cobbie Cuddles." The 10" pot belly bear on the right was offered in 1983. Body tag: "Dae Kor Made in Korea." Paper tag: "Cobbie Cuddle Soft As A Hug." *Courtesy of Ginny Kreitler.*

COBBIE CUDDLERS ® (shoes) - Undressed 9" seated light brown plush bear with tan snout and ear linings, plastic eyes and brown plastic nose, floss mouth, cheek dimples, raised arms. Tush tag: Seated bear logo "Cobbie//Cuddlers®" Tag reverse: "Made in Korea." Second tag: "© 1988//H&T Marketing Co.//All Rights Reserved." $8-$15.

Coca Cola

Coca Cola and its advertising polar bear are familiar worldwide. We all fell in love with him in television commercials during the Olympics. However, other types of bears have also been used in advertising the carbonated beverage, which was first concocted by a Georgia pharmacist in 1880 as a remedy for upset stomach, nervousness, or headache. By 1887 the script Coca-Cola logo was trademarked and in use.

COCA-COLA/WHATABURGER - 8" seated pale yellow plush bear with beige furry ears, tan snout and back paw pads, plastic eyes, and chocolate brown plastic nose. Green cotton knit T-shirt is part of his body. Shirt says (in white): "Enjoy Coca-Cola, Trademark ®." Leg tag: "Whatabear...g-r-r-r-™" Tag reverse: "© CALTOY, Inc., Los Angeles, Handcrafted in Korea." (This company is not to be confused with WHATABEAR of Seal Beach, California.) This bear was a promotional item for Coke and Whataburger (similar to Burger King) in the late 1980s in Pensacola, Florida and San Antonio, Texas. $20-$25

COCA-COLA - 10" seated gray bear with white snout, ear linings, and back paw pads, brown velour nose, pink flannel tongue sewn down, cord loop sewn to back of head, painted plastic hooded oval side-glancing eyes, wears sunsuit of red and white Coca-Cola print, pants are part of the body, bib and straps are free, holds a red satin Coca-Cola "can" with metallic silver satin top and bottom, straw replaced. Red paper ear tag: "Authorized Product, The Coca-Cola Company" Paper tag reverse says: "Distributed by// Play-By-Play// Toys & Novelties, //San Antonio, Texas// © 1993 The Coca-Cola Company// Made in China." Back tag: "Coca-Cola Plush Collection." Coca-Cola has been using the script "Coca-Cola" logo since 1887. $12-$16.

COCA-COLA - 9" brown plush pair of Coca-Cola sipping soda bears, dressed in red and white Coke print outfits, sharing a Coke. Called "Loving Coca-Cola® Couple" and "Beary Nice Bears." Advertised in mail order gift catalog in Fall 1994, price for the two bears $27.95.

COCA-COLA - 1994 "The Limited Edition Coca-Cola ® Santa Heirloom Collector Doll" from Franklin Heirloom Dolls has a tiny beige bear perched atop Santa's sack. The bear has a white snout and white ear linings. Santa is 21", porcelain, dressed in red velvet and faux fur, shiny boots, and brass belt buckle. The hardwood display base has a specially minted Coca-Cola medallion imbedded in it. Santa holds a bottle of Coke in one hand and the bag of toys in the other. The doll was inspired by the illustrations of artist Haddon Sundblom, first drawn in 1931 for Coca-Cola. Edition limited to 45 firing days. Issue price $195.

Close up of little bear in Heirloom Santa's sack.

Above: COCA-COLA - 1996 "Cubby, The Coca-Cola Heirloom Collector Teddy Bear" (first-ever Coca-Cola authorized heirloom collector Teddy Bear).Cubby is 19" handcrafted of genuine mohair, fully jointed, with hand-set glass eyes, hand-embroidered face, and hand numbered. He's dressed in blue pants, a plaid flannel shirt, and red billed cap with the Coca-Cola name in white. He comes with a red Coca-Cola wagon "just like the ones little boys created from real "Coca-Cola" crates years ago." In his paw is a golden Mint Mark medal, stating his special status as a Coca-Cola Heirloom Bear. Made by Franklin Heirloom Bears. Issue price $195.

Left: COCA-COLA - 1995 "Megan, A Limited Edition Coca-Cola ® Heirloom Collector Doll" from Franklin Heirloom Dolls, authorized by Coca-Cola ®. Megan is 19", crafted of hand painted porcelain with inset eyes, dressed in her turn-of-the-century (1900) best. She pulls a wooden wagon "which Daddy made for her from an old "Coca-Cola" crate. It's the perfect place for her new best friend, Teddy which she just bought at the toy store." Teddy is jointed, beige plush, with a maroon neck ribbon. Edition limited to 45 firing days. Issue price $195.

COCA-COLA - 1998 "Bonnie, A Coca-Cola Heirloom Collector Teddy Bear." Bonnie is 18" handcrafted of genuine mohair, fully jointed and poseable, with hand-embroidered features, handset glass eyes, sueded paw pads, and is hand numbered. She has a golden Coca Cola medallion on her shoe. Her lace dress and hat of Grandma's were "found in the attic." She comes with her own Coca-Cola trunk. The lid opens to reveal her treasures. Made by Franklin Heirloom Bears for Coca-Cola. Issue price $195.

COCA-COLA - Brand new in 1998, 20" "The Coca-Cola Heritage Soda Fountain Bears" Plush bear wears vintage-look white shirt and bow tie, red/white striped vest, black striped trousers, and a soda "jerk" style hat. He holds a glass with holder in his left hand. His left foot says "Coca-Cola" in script.

COCA-COLA - 8.5" seated white Coca-Cola polar bear, plastic eyes and nose, black leatherette back paws, holds 3" plastic Coke bottle. Tush tag: "Coca-Cola" Plush Collection, Shell Made in China, © 1993 Coca-Cola Co., all rights reserved." Paper tag: "Always Coca-Cola// Coca-Cola ® brand Plush Collection." Reverse: "... Distrib by Play-By-Play Toys & Novelties// San Antonio, TX © 1993. Purchased at the Coca-Cola factory store in Atlanta, Georgia. Note the slender pointed snout. Fabric feels more like flannel than plush. The 10" version of this bear sold in a 1995 catalog for $14.99. *Courtesy of Nancy Vanselow.*

COCA-COLA - 7" chubby white plush Coca-Cola polar bear, plastic chest button says "Always" in green above a red circle with Coke bottle and "Coca-Cola" in white. Tush tag is dated © 1993, says "Made in China." His folder-like paper tag has a polar bear on the cover, and inside says: "The Bear Everyone is Thirsting After! You've seen him on TV delighting fans in millions of dens across the country. This is the one and only Coca-Cola Polar Bear. A lovable visitor from the frozen north, he's guaranteed to warm the home of any family he joins." Paper tag back: "© 1993 The Coca-Cola Company, Distributed by Cavanaugh Group International, Made in China." In 1993/94 this bear was used as a premium by Amoco Oil Company gas stations, and Hardee's restaurant chain. On the left a 7" bear, more slender, with a longer thinner snout. A 3" Coke bottle is sewn to his right arm. .Fabric feels like a cross between felt and flannel. Tush tag is different. Front says: "Coca-Cola ® brand Plush Collection" and reverse: "Made in China" and "© 1993 Coca-Cola." $10-$15 each.

COCA-COLA - On the left: 6" seated white flannel-like bean bag polar bear, holding a 1.5" plastic Coke bottle, plastic eyes and nose, floss mouth, small flat tail. Dressed in a red and green Coca-Cola bottle and bottle cap print shirt that is part of his body. Paper tag resembles silver bottle cap, inside: "Coca-Cola Polar Bear in Argyle Shirt." Leg tag: " © 1997." Currently available $10-$15. In the center: 6" seated white flannel bean bag polar bear holding plastic 1.5" Coke bottle. Plastic eyes and nose, red ribbon bow tie. Green cotton visored old-Texaco-style hat has red plastic button with Coke bottle and "Coca Cola" on it. Paper bottle cap shaped ear tag reads: "Collectible Coca-Cola Bean Bag Plush" Inside the paper tag: "Coca-Cola Polar Bear in Drivers Cap and Bow tie" and "Coca-Cola Brand Plush, (c) 1997 The Coca-Cola Company." Leg tag: "Coca Cola.... Made in China" Currently available $10-$15. On the right: 6" seated white flannel-like bean bag polar bear, holding a 1.5" Coke bottle. Nose tipped up in more friendly fashion. Wears red knit sweater with green neck, cuff, and bottom trim, and "Coca-Cola ®" machine embroidered on the front. Same tags as his companions. Various of these bears were accessorised differently, with other shirts, pink hair ribbons, and the like. Currently available $10-$15.

COCA-COLA - 11" seated Coca Cola polar bear with 10" leg spread, flat tail. Chest button does not say "Always." Tag "© 1993." The Spring 1995 Coca-Cola Catalog offered bears with these chest buttons, holding Coke bottles, in three sizes: 10" ($15), 13.5" ($35), and 30" ($185). *Bear courtesy of 2nd Chance Home Furnishings.*

COCA-COLA - 12" white plush Coca-Cola polar bear wearing velvet Santa hat (tassel to the left) with "Coca-Cola" machine embroidered in red on its rim, and 1.5" chest button. Body and paper tags both read: "© 1996." Original Montgomery Ward price tag says $9.99. In 1995 an animated plush Coca-Cola Santa with a large chest button, holding a Coke bottle in his left hand, and with his "Coca-Cola" embroidered hat rim and hat tassel to the right side, was offered for $44.99.

COCA-COLA - Two sizes of Coca-Cola polar bears never removed from their boxes. On the left 12" bear has a closely sheared muzzle, wears a red scarf with white yarn fringe and "Coca-Cola" stenciled on it in white. He holds a 4.5" plastic Coke bottle. The back of his box says: "Distributed by Play-By-Play Toys & Novelties, San Antonio, Texas, © 1993 The Coca-Cola Company, Made in China." Original Wal-Mart price $11.97. On the right 8" white bear of shorter plush and less elongated muzzle. Same markings on box back. Holds 3" Coke bottle. Original grocery store price $7.99. Neither bear has a chest button. In 1997 a discount gift catalog showed a 9.5" Coca-Cola polar bear with no chest button, seated on a flat red cardboard display box, and holding a 4.5" plastic Coke bottle. His price was $9.99. Current values in their boxes: 12" $25, 8" $20, 9.5" $15.

Walgreen's advertised, in 1995, this Coca-Cola Santa whose hat has faux fur rim (no embroidery, tassel to the left), chest button, and red bow tie. In 1996 and 1997 Walgreen's Coca-Cola Santa bears had plain hat rims, with tassels to the left, but no bow ties, price $9.99. Walgreen's ads in 1996-97 also pictured a 10" standing Coca-Cola polar bear with a red and green bow tie. Mini plush Coca-Cola Santa bear ornaments, with Santa holding a tiny Coke bottle, could be purchased for $2.99 during a special promotion.

Above: COCA-COLA - 8" chubby plush Coca-Cola polar bear's chest button says "Always" above the Coca-Cola (nickel sized) red circle. The 12" bear's button does not say "Always." Both wear their original paper tags. ©1993. In November, 1994, Target Stores advertised a 10" bear wearing the non-Always chest button who had "1994" embroidered in white on his black left paw pad. "Offer good while supplies last." Sale price $8.49. Bet they went fast. Most likely it was a Target exclusive. Current value $15 up for each.

Right: COCA-COLA - 11" white polar bear wears a red T-shirt with the design of a white outlined red heart and arrow and a banner that says "Coca Cola." T-shirt is cotton/polyester blend and imported. He appeared in the Spring, 1995 Coca-Cola catalog, as a catalog exclusive, called "Valentine 'Coca-Cola' Polar Bear," priced at $24.

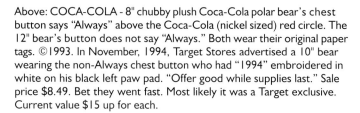

A 13.5" Coca-Cola Bear with a box full of little 6" Coca-Cola Bears. All are wearing red scarves with white fringe and the white Coca-Cola script signature. None have chest buttons or Coke bottles. They were offered in a gift catalog in 1995. Original prices 13.5" $30.75, 6" $10.75, and a 30" (not shown here) $185.

COCOS (restaurant) - 12" golden brown plush bear with cream snout, ear linings, and back paw pads, plastic eyes, black plastic nose, red flannel lined mouth. Looks like "Shoney" bear except wears kelly green shirt with "COCOS" in white across shirt front with his dark blue denim jeans with shoulder straps. Shirt has Velcro® back closure. No tags. $5-$10

Above: CRACKER JACK® - Two piece set: 16" Cracker Jack® and 7" Bingo, limited edition 1,000. Golden bear in sailor suit and hat saluting with his left hand and holding a miniature box of Cracker Jack® in the right hand. Made by Cooperstown Bears, Buffalo Grove, Illinois as part of their "Americana" collection. Appears in their 1998 catalog. The Cracker Jack confection was concocted in 1890, the sailor boy symbol came into use just prior to World War I. In 1976 Mattel made a boxed Cracker Jack® advertising doll and named him Sailor Jack.

Right: CRACKER JACK ® - 15" cream distressed curl plush Cracker Jack sailor bear with his 5" dog Bingo, new in 1998 from Douglas Cuddle Toys as part of their Company Classics collection. Jack salutes with his left hand and holds a miniature box of Cracker Jack (his favorite sweet snack) in his right hand. His white "Dixie cup" sailor hat has his name on it in red. Limited edition, individually numbered and boxed. Issue price, 1998, $110. A 5" Mini Cracker Jack bear was added to the line in 1998, Issue price $16.50

CRACKER JACK® - 16" dark brown curly fur, fully jointed, Cracker Jack bear wears only a sailor hat, and sailor middy collar with red tie. He gives a snappy salute, emulating the original advertising logo on the old Cracker Jack box. Made by North American Bear in 1993. Issue price $48.50.

CRAYOLA™ - 19"Crayola bears were sold in gift and department stores in 1986. They came in a rainbow selection of colors: green, orange, purple, yellow, red, blue. Smaller (8") bears were available in the same colors from Burger King in 1986 with purchase of a fast food item. A different color was available each week. They were also advertised in newspaper Sunday advertising supplements free with proofs of purchase. The 19" green bear and 8" orange bear above have the same tag: "© 1986 Crayola, Binney & Smith Inc, Manufactured for Graphics International Inc, K.C. Mo" The green bear was made in Korea and the orange bear in China. The other 8" bears pictured have tags that delete the "Binney and Smith" and just say "Graphics International." "Binney and Smith" probably predate the plain "Graphics International" bears. The yellow and red (most commonly found colors) were made in Korea; the blue and purple were created in China. The red and yellow bears wear their paper "Burger King, Happy Holidays" tag. Current value 19" $25, 8" with paper, shirt and body tags $10-$15, without tags $5-$8.

CRAYOLA™ - 19" yellow Crayola bear with 10" pink Crayola bunny. Bear's sweater is removable, as is bunny's T-shirt. White and red plastic "Heartline" tag in bear's left ear. Bear's leg tag: "© 1986 Crayola, Binney and Smith Inc., Manufactured for Graphics International Inc, K.C. Mo 64141, Made in Korea." Bunny's tag is the same as Bear's, except for an added separate "Heartline" leg tag.

CRAYOLA™ - A pair of 8" Crayola bears, purchased at Burger King, wearing their Burger King paper tags. Both were made in China, 1986. The Crayola company began in 1903 with a box of 3 crayons in primary colors. In February 1997, Crayola celebrated the 40th anniversary of their "Crayola 64" crayons. An original (1957) box of the crayons was donated to the Smithsonian Institution. *Courtesy of Oneida O.Callaway. Photo by Mary Ann Callaway Dennis.*

CRUNCH 'N MUNCH (caramel popcorn snack) - 8" beige bear with white pads, snout and ear linings, yellow knitted scarf says "Crunch 'n Munch" in red on the right. "My Little Toffee" Bear offer, 1986, $3.50 plus 2 UPCs. Tag: "Atlanta Novelty, Inc., Subsidiary of Gerber Products Co, Woonsocket, RI 02895." Gerber sold Atlanta Novelty in 1988. *Courtesy of Jean Laughery, Photo by Vivian L. Gery.*

CURITY ® (first aid supplies) - (1) 11" seated bear, cinnamon brown, beige snout, plastic eyes, black floss nose, red felt mouth sewn to lower mouth flap, white flannel diaper fastens with white metal snaps and says: "Curity ®" in green across the front. Red plastic heart glued to chest says "Dydee Bear." Leg tag has two circus tents logo and "Animal Fair, Eden Valley, MN 55329" Tag reverse: "© Animal Fair, Sewn in Haiti." (2) 11" bear same as (1) except wears no diaper, glued red chest heart has printed "dy-dee bear ®" suspended from a printed safety pin and tag reads: "Animal Fair, Chanhassen, Minnesota 55317. Tag reverse gives no country of manufacture. Circa 1950s-1960s. $30-$45 each.

D

D.A.Y SPORTSWEAR - 24" white shaggy plush bear with blue and white striped ear linings and back paw pads, large black plastic nose, plastic eyes, floss mouth, wearing removable one piece outfit with sailor collar. Blue and white striped pants and bow tie, navy blue top has "D.A.Y. Navy" machine embroidered in red script. Red stars on attached white drawstring belt. Plastic heart shaped tag on bear: "The// Reinhart// Collection." Tag on outfit:."D.A.Y//. Sportswear// Baby." Second tag: "100% Cotton//size 18m.....Made in Philippines." Flintstone print underpants may have been added. $8-$20

DEL MONTE (foods) - 9" seated, 11" standing, light brown plush bear with tan snout and ear linings, plastic eyes, dark brown velour nose, floss mouth, tan flannel brimmed hat sewn to head, dressed in red and white checked shirt with collar and back Velcro ® fastening, dark blue denim overalls with leg cuffs, two rows of white stitching at waist, red Del Monte patch on the overalls front and opening for tail in the back.. Leg tag: "Del Monte// Pumkin™ The Brawny Bear//© Del Monte Corp - 1985," Leg tag reverse: "R. Dakin & Company….Product of Korea," Paper ear tag: "The Story of Brawny Bear." Some collectors refer to him as "Del Monte Farmer Bear." He was the last in a series of 10 items offered free with 75 Del Monte labels. (A case of Del Monte vegetables for Christmas completed one collector's set.) Another collector purchased her "Brawny" through a magazine ad. *Courtesy of Ginny Kreitler.* $25-$35.

DILLARDS (department store) - 14" seated white plush bear, with plastic eyes and nose, wears red knit sweater with green trees and the date "1987" in green. Hat has white pompom and ear holes. Leg tag: "Made in Korea." A Christmas promotion. *Courtesy of Ginny Kreitler.*

DIAL SOAP - 10.5" white plush sitting bruin with red felt scarf. This teddy is whisper soft with huge footprints. Tag: "Friendly Teddy, Russ, Berrie Co, Inc, Oakland, NJ." "Friendly Teddy" was offered by Dial Soap in 1985 for $9.95 and 4 Dial Soap wrappers. Offer expired December 31, 1985. *Courtesy of Jean Laughery, Photo by Vivian L. Gery.*

DISNEY -1991 Fourth Walt Disney World Teddy Bear Convention, 12" black mohair bear with hump back and growler, white suede paw pads, black button eyes, black stitched nose, brown stitched toes and mouth, Black and white Mickey Mouse mask included. Red neck ribbon streamers say: "Walt Disney World's Teddy Bear Convention 1991" and "Steiff, Germany." Edition limited to 1,500 pieces. Original price $175. *Courtesy of Nancy Vanselow.* 1998 value $800.

DOMINO ® SUGAR - 9" sitting brown bear with white snout, ears, and paw pads; plastic eyes and nose, floss mouth. Yellow shirt says "I (red heart) Domino ®" Tagged: "Russ." Was originally purchased through a magazine ad. *Courtesy of Oneida Callaway. Photo by Mary Ann Callaway Dennis.*

Domino Sugar box that comes out of the back of the bear.

Domino Sugar box reverses around the bear and becomes a sleeping bag.

DOMINO ® SUGAR - 10" seated all white plush bear, plastic eyes, black velour nose, wears yellow cotton chef's hat with a blue "D" and yellow cotton apron with "Domino ®" (in blue) and "Sugar" in red. Bear zips down the back to reveal a satin replica of a box of sugar, marked "Domino ® pure cane Granulated Sugar." The bear's legs and body tuck into the satin box which becomes a sleeping bag. Leg tag: "Commonwealth, Toy and Novelty Co., NY, NY; © 1988, Made in China." The 1989 promotion was called "Bear In A Bag", original price $8.95. An especially well made and intriguing bear. $20-$35.

TED DREWES (frozen custard, St. Louis) - The 11.5" fully jointed brown bear on the right wears a removable shirt with "Ted Drewes Since (cone shape) 1934, Frozen Custard, St. Louis." Bear was available in 1986. Paper ear tag: (crown shape) "Princess." The light tan unjointed 13.5" standing bear on the left was offered in 1987. *Courtesy of Ginny Kreitler.*

EAGLE FAMILY DISCOUNT STORES - 12" seated peach plush bear with tail, white snout, ear linings, and back paw pads, plastic eyes, black plastic nose with indented nostrils, red felt tongue. Tag on bottom has silhouette of an eagle "EAGLE// Manufactured Exclusively For// EAGLE FAMILY DISCOUNT STORES INC., Opa-Locka, Florida 33054", Tag reverse: "Made in Korea." $10-$18

DUSTBUSTER/ LA-Z-Y BOY - On left 9" standing brown Dustbuster Bear with red neck bow is tagged on his left foot "Dusty" and his white apron says in red "I love my Dustbuster." At right the 7" gold koala bear came with a La-z-y Boy recliner chair purchase. Maroon shirt, which is part of body, says: "La-Z-Boy™". *Courtesy of Ginny Kreitler.*

ECKERxD

ECKERxD ("America's Family Drug Store") - 16" beige seated plush bear. Dressed in blue knit cap and scarf with big white pompoms. Cap has "Humfrey" knitted into the front in white; scarf ends have knitted white "H" on each. No body tags. Folder type paper tag front: "The Bear with the Magical Hugs!, Humfrey Hug-A-Bear" and drawing of a little girl holding Humfrey in his cap and scarf. Inside the tag: "The Story of Humfrey. Humfrey Hug-A-Bear is my name, And magical huggin' is my game!, At Santa's shop, on my birthday, I longed to be "special" somehow, someway. 'Oh, make me special, please, Santa, please? So I will be noticed there under the tree! But Santa knew best, he knew what I sought, He knew what I wanted just couldn't be bought. He said, 'From now on, whomever it pleases Will feel good all over when they give you squeezes! This gift is far better than deeds you might do, It's my secret gift and I give it to you!' And what's the secret about my huggin'? I learned a lesson 'bout giving out lovin' It's not so important to be a star, My hug means, I love you, just the way you are!" Back of tag shows Humfrey under the Christmas tree and "ECKERxD ® America's Family Drug Store." Although undated, I believe this is the first Humfrey because the outfit matches the original 1987 Eckerd storybook which introduced Humfrey Hug-a-bear. The 8 page paperback is pictured. It's © 1987 Jack Eckerd Corporation, written by Karen Sebourn. Original book price 69 cents. Bear with paper tag and book *Courtesy of Muriel Hoffman.* $25-$35.

ECKERxD - 8" beige "Humfrey" bear in a red bag with his name on it. (An unusual size for Humfrey.) Nightshirt and nightcap have red and green wide a part stripes on a white background. The back of the bag says: "Humfrey is my name, and magical hugs is my game." The story tells how Humfrey asked Santa to make him special. The secret of his magical hugs is "I love you just the way you are." Purchased on-site at Eckerd drug store. The nightshirt print is the same as 16" Humfrey; possibly made the same year. *Courtesy of Oneida Callaway, Photo by Mary Ann Callaway Dennis.*

ECKERxD - 16" seated beige plush bear, in removable white/green/red striped flannel nightshirt and nightcap with red and green yarn tassel. Nightshirt has "Humfrey" machine embroidered in red above a brown embroidered paw print. Same size and face as Santa and Peek-A-Boo Humfreys. Tush tag: "© MTY International Co Ltd, Made in Taiwan, R OC." *Courtesy of Muriel Hoffman.* $15-$20.

ECKERxD - 16" seated beige plush bear with white snout, paw pads, and ear linings, plastic eyes, black plastic nose, stitched mouth and stitched red tongue. He wears a red velvet neck bow, red velvet Santa hat with faux fur brim and tassel. A red velvet heart, stitched to his chest, says "Humfrey 1989" in gold thread. A back zipper allows access to his power pack. When batteries are inserted it activates an audible "heart beat." Tush tag: "© MTY International Co, Ltd, Made in Taiwan, ROC." $20-$30.

ECKERxD - 16" seated beige plush bear. This Humfrey's only accessory is a blue print stuffed bow tie. Magnets in each paw and on either side of nose allow him to play peek-a-boo. On left foot is a yellow appliqued bear and the legend "Peek-A-Boo Humfrey, '92" in red. Tush tag: "©Dan-Dee International Limited, Jersey City, NJ 07305, Made in China." Eckerd's parent corporation is the J.C. Penney Co. $15-$25.

F

FAMOUS BARR DEPARTMENT STORE (St. Louis) - 17" white bear sold at Christmas 1986. His red knit hat with holes for ears says "Famous Bear" in white. Red scarf has a snowflake and evergreen trees design. The 18" Paddington Bear with light blue hat, red felt coat, and yellow boots was used by the store in a Christmas advertisement in 1974. *Courtesy of Ginny Kreitler.*

FABRI-CENTERS - 5" beige bear with jointed arms and legs and stationary head, black plastic eyes, black floss nose, beige felt paw pads, single floss toe marking on each paw, red neck ribbon. Side tag: "Manufactured for Fabri-Centers of America", reverse: "Made in China." $2-$4

FORENZA (Italian line of women's clothes) - 14.5" white bear "Paolo", 1986, is said to represent "the Italian way of life, to see with the eyes of a child." Paolo wears red shorts, purple/navy top with red wrist bands and white "Forenza" on shirt. Plastic eyes and nose. *Courtesy of Ginny Kreitler.*

FLORIDA FEDERAL (bank) - 9" black and white plush panda, white tail, plastic eyes with black felt eye surrounds, black plastic nose, black floss mouth, wears navy blue sleeveless sweatshirt. Shirt has "I (heart) Florida Federal" in white on the chest. Leg tag: "Logo Bear © 1985" Tag reverse "Made in Korea" $5-$8.

FRANCO AMERICAN ® (parent company: Campbell's ® Soup) - 18" bright rust colored plush, definitive thumbs on forepaws and toes on back paws, big tail. Button: "I Love Teddy O's Original." Tag: "Teddy O's, Case/ Dunlap Plush Promotions, California." In 1990 "Teddy O's Bear" was a premium, price $6.95 and 2 front labels from Sporty O's, Teddy O's, or Spaghetti O's. *Courtesy of Jean Laughery, Photo by Vivian L. Gery.*

FRIGIDAIRE (refrigerators) - 18" honey brown shaggy pile bear, white snout, plastic eyes, black plastic nose, black floss mouth, wears blue and white horizontally striped polyester T-shirt with "Frigi Bear" across its front. Tag at back waist: "Custom Designed for Frigidaire" Tag reverse: "100% acrylic, Made in Taiwan R.O.C." Loosely stuffed, floppy and huggable. Used 1988-89 in Florida as a product promotional item, purchased in Pensacola. (Frigi-Bear was author's first advertising bear, thus is the catalyst behind this book.) $10-20.

FRIGIDAIRE - Another version of Frigi Bear in white plush with pink plastic nose, and pink removable shirt that says "Frigi Bear." *Courtesy of George B. Black, Jr., Teddy Bear Museum of Naples, Florida.*

FTD - 6" tan wooly bear, plastic eyes, velvet nose, Velcro ® on front paws, blue morning glory is sewed to foot, straw hat with light blue ribbon, red satin neck ribbon. Tush tag: "Made exclusively for FTD by// Mohair Bear Co.// Salt Lake City, UT 84054// Made in Bangladesh." Materials tag is in English and French. *Courtesy of Nancy Vanselow.*

Front view of 8" Raikes-like FTD bear beside his original 8.5" plastic hollow tree flower container. *Courtesy of the Vivian Vanselow collection.*

FTD (Florists Transworld Delivery) - 8" seated wooden faced bear with wood back paw pads, red satin vest. "FTD Collector's Series// © FTD, Inc., Southfield, MI, Made in China ... 100% polyester fiber stuffing// not a toy." Tag reverse in English and French. Bear's 8.5" plastic hollow tree flower container has carved (incised) heart. Bear came fastened to side of tree. *Courtesy of the Vivian Vanselow collection.* $35-$50.

FTD - 10.5"all white plush bear with plastic eyes, black plastic nose, white satin shirt front with red velour lapels and bow tie is attached to body with elastic around the neck and stitches at waist. Back tag: "Made Expressly For FTD By//Dan Dee International Limited//Jersey City, NJ 07305//Made in China// (c) 1990 FTD, South field, MI 48037// Surface washable with damp cloth." Accompanied a floral arrangement. $5-$12

FTD - 10" seated pink plush bear with slightly chubby tummy, light blue heart print ear linings and back foot pads, plastic eyes, white snout, pink plastic nose, pink floss mouth, neck bow of white ribbon with pink and blue flowers and green leaves on it. Tush tag: "Made Expressly for FTD by Dan-Dee International Limited, Jersey City, NJ 07305, Made in China, © 1992 FTD ® Southfield, MI 48037" Probably accompanied flowers sent to a new mother. Separate contents tag is written in English and French. $10-$15.

FTD - Five 6" peachy beige plush bears with tail, white pointed snout, back paw pads and ear linings, body facing right with bear looking over his shoulder, plastic eyes, brown plastic nose, brown floss mouth, paws held together by Velcro®. Leg tag (on 3 bears on left): "Made Expressly For FTD By Duk Seong Trading Co Ltd, Seoul, Korea, © 1991 FTD ®, Southfield, MI 48037." Leg tag on chubbier bears (on right): "Made Exclusively For FTD ® By Kolob Intertrade Co., Ltd, Bangkok, Thailand, © FTD, Southfield, MI 48037" (no date given). Thailand bears have less pointed noses and are more softly stuffed. Separate contents tag is in English and French for all bears shown. A 7" lookalike bear, clutching a clear glass vase with a single blue carnation, was offered in a 1995 mail order catalog priced at $16.75.Current, as is, $5-$8 each.

FUJI FILM - 17" brown plush "Fuji Holiday Bear", with "safety lock eyes," is dressed in red cap with green trees and red muffler fringed in green with "Fuji Film" spelled out in white. The bear was free with 15 proofs of purchase of Fuji Film or 5 proofs of purchase of Fujicolor QuickSnap Cameras. Or for $9.95 and 3 proofs of film or 1 proof of camera purchase. Offer expired December 1, 1992. $10-$15

G

GARST SEEDS (corn seeds) - 16.5" colorful bear, orange with yellow belly, round head and round body, "Garst" in yellow against an orange rectangle on his tummy. Tag on right foot: "Paw Pals, 1984." *Courtesy of Ginny Kreitler.*

THE GAP (clothing stores) - 7" red flannel bear with outstretched arms, black floss eyes, nose and mouth, red plaid neck ribbon, rattle inside. Back tag: "Baby GAP, The GAP, 1 Harrison St, San Francisco, CA 94105, Made in China." Tag reverse gives washing instructions. $6-$10

GENERAL FOODS CO. (Post Cereal Division) - 15" brown plush bear with "Sugar Bear" in white on his removable high-waisted blue shirt which exposes his bare belly. Tag: "Animal Fair, Inc., Minneapolis, MN 55440." This was a mail in offer in 1989, price $8.95 and 2 UPC symbols from Super Golden Crisp Cereal. *Courtesy of Jean Laughery, Photo by Vivian L. Gery.*

GENERAL FOODS CORP. (cereal, etc.) - A quartet of "Sugar Bears." Bear second from left is 5", the others are 4.5". Two bears at left are of flannel; felt features are glued on, and blue knit shirts (with a pattern of the name "Sugar Bear" in black) are removable. Bears at left are from 1987 and predate bears at right. They were free enclosed in specially marked boxes of Super Golden Crisp wheat puffs. Two bears at right are of cotton fabric printed front and back. Bear in patterned shorts was called "surfer." Tags are the same for all but "Santa." Santa's tag reads: "© 1990 Kraft General Foods Inc." while the other tags are undated and say: "General Foods Corp, White Plains, NY 10625, Made in China." Santa has loop on head for use as Christmas tree ornament. He was musical, played Jingle Bells, Rudolph, We wish you a Merry Christmas. Santa bear $15-$20, others $5-$9. *Promotional information courtesy of Jean Laughery.*

GENERAL MILLS - Three look-a-like 8.5" bears. On left he wears a red shirt with "General Mills, I (heart) Brownies" Tag: "Logo Bear, Roly Poly Bear." In center: ELECTONE INC. (hearing aids) bear from 1986 in blue shirt with white letters. On right AMOCO MOTOR CLUB in red shirt was free in 1986 with purchase of an Amoco Auto Club membership. *Courtesy of Ginny Kreitler.*

GENERAL FOODS CO. (Post Cereal Division) - 16" brown plush "Sugar Bear" with name spelled out in white letters on blue shirt which is part of his body. Offered in 1977-79 for $4.75 plus two labels from Post Sugar Crisp cereal. Offer expired July, 1979. Tiny 4.5" bear from 1986 wears removable shirt with "Sugar Bear" printed all over it. *Courtesy of Ginny Kreitler.*

Right: GENERIC PRODUCTS (plain labels) - Light yellow cotton 10.5" seated bear with a Universal Product Code on his right foot, small stitched eyes, mouth, and nose. Chest says "Basic All Purpose Durable No Frills Bear Lost in a Designer World" "Plain wrap product." Attached paper tag reads "When you could care less about having the very best." *Courtesy of Ginny Kreitler.*

Gerber Products

GERBER PRODUCTS - 8" seated red bear, with white tummy, snout and ear linings, plastic eyes and nose, open mouth has red bottom lining. This "message" bear holds a lace-trimmed satin heart that asks "Guess who luvs you?" His tag is the same as beige bear except "Made in Taiwan." $6-$10.

GERBER PRODUCTS CO. (baby food) - 13" fully jointed beige bear with white snout and cream felt paw pads, amber plastic eyes, black floss nose and mouth. Back seam tag: "Atlanta Novelty, Division of Gerber Products Co., New York, NY 11101, Made in Korea." Gerber sold Atlanta Novelty in 1988. Ice cream cone is not original to bear. $10-$15.

Above: GERBER PRODUCTS - 5" seated black and white plush panda, plastic eyes, red neck ribbon. Paper tag: "Gerber// Cuddlies// Made in Taiwan." Side seam tag: "Gerber Products Co.// Fremont, Michigan 49412, Made in Taiwan, ROC." *Courtesy of Shirley Taylor.*

Left: GERBER PRODUCTS - 8" standing brown bear with arms tacked to body, has an appealing look. Leg tag: "Gerber Products Co, Fremont, Michigan 49412, Made in Korea." Washing instructions are on tag reverse. Gerber's motto is: "Babies are our business, our only business." In addition to baby food, their products include baby furniture, bibs, bottles, rattles, toys, and clothing. Gerber's main office and food plant are in Fremont, Michigan near Muskegon. The firm celebrated its 70th anniversary in 1998. Bear $5-$8

GERBER PRODUCTS - 7" light beige plush koala, black plastic nose and eyes, floss mouth, red neck ribbon. White ears, tummy and back paw pads. Leg tag: "Gerber Products Co// Fremont, Michigan 49412, polyester, Made in Korea" Washing instructions on tag reverse. Paper tag: (Gerber baby face, ® logo) "Gerber// Cuddlies ™". *Courtesy of Peggy Monahan.*

GERBER PRODUCTS - 10" brown terry cloth "learn to dress" bear, machine embroidered eyes, nose, and mouth. Blue shirt, red pants, and beige shoes are part of the body. Yellow suspenders snap, right shoe ties, left shoe fastens with Velcro®, back of shirt has a zipper, and green hat Velcro® to back of head. A cord fastens hat to body to eliminate loss. Body tag: "® Gerber Products Co., Fremont, Michigan 49413, Made in China, © 1994 Gerber Products Company." Paper package tag: "Gerber Graduates Activity Doll, helps introduce self-dressing skills, Totally Tots™, Made in China to Gerber specifications." Gerber Graduates is food for toddlers. Big Lots price for mint-in-package bear in 1998 was $1.00. Current price $6-$10.

GERBER PRODUCTS - 19.5" brown plush bear, tan snout and paw pads, red velvet bow tie, white plastic disk tag (in ear or on bow tie) says "Gerber Tender Loving Care Bear." Bear came in 3 color choices: brown, tan, or white. This 1987 promotion required answers to a baby food quiz and $10. Retail value in 1987 $24.95. *Information courtesy of Jean Laughery, Photo by Mary Ann Callaway Dennis courtesy of Oneida Callaway.*

Above: GETTY OIL CO. - 11" brown plush bear wearing a removable T-shirt with red sleeves and "I'm a Getty Teddy" in red on the front. Tag: "Made Especially for Getty, 1986, KBR Group, Calif. USA." The 1986 holiday promotion offered customers Teddy for $3.00 with the purchase of 8 gallons of gasoline. *Courtesy of Jean Laughery, Photo by Vivian L. Gery.*

Above: Center: GETTY OIL CO. - 11" beige plush "I'm a Getty Teddy" bear with same description and tags, just a lighter color. *Courtesy of Shirley Taylor.*

Above Right: GETTY OIL CO./AERO SERVICE CO. (heating oil) - 11" tan plush bear dressed in warm-up suit with red pants, white shirt with "Aero" in red, white headband, and white terry towel around his neck. Tag: "Made Especially for Getty, 1986, KBR Group, California, USA." In 1986 the "Warm-up Teddy" promotion was a free bear given to first time customers. *Courtesy of Jean Laughery, Photo by Vivian L. Gery.*

GODIVA CHOCOLATES - 6" beige bear with jointed arms and legs, plastic eyes, floss nose, mouth, and paw claws, tan felt paw pads, red ribbon neck bow, gold cord loop on head. A bright pink foil wrapped chocolate heart (2" high, 2-1/2" wide) is fastened to his chest with gold cord. Arm tag: "Creative Marketing// Concepts Inc//Cleveland, OH 44131// Made in China" A metallic gold paper tag pictures Lady Godiva on horseback and reads: "GODIVA Chocolatier// Milk Chocolate// Chocolat Au Lait// Net Wt 1 oz" The tag's reverse lists ingredients and says "Made in U.S.A." With tags and chocolate $15-$20

GIBSON GREETINGS INC. (greeting cards) - 8" seated bear with beige silky shaggy fur, beige sheared snout, oval plastic eyes, black velour nose, black floss mouth, brown felt paw print on each paw pad, no tail. White cotton shirt is part of bear's body, red velveteen vest, black velvet bow tie, gold lamé heart shaped pocket "watch" on gold cord "chain." Paper ear tag: "Debonair Bear" with black top hat and cane and "Guaranteed to Warm the Heart." Paper tag reverse: "Manufactured for and Distributed by © G.G.I., PO Box 145485, Cincinnati, OH 45327, © 1998 G.G.I., Made in China." Tush tag says the same except it spells out G.G.I. as "Gibson Greetings Inc." Has a pleading, take-me-home face. Bear sold in grocery store, original price $13.99. Current $15-$18.

GORHAM (silver) - 19" undressed all white plush bear with 2.5" black nose and dime-sized plastic eyes, large feet and beaver-like flat tail, straight back and chubby tummy. White satin tag under tail is printed in pink: "Cuddly Companions™ // (c) 1986 All Rights Reserved// GORHAM// Providence, RI 02907." Tag reverse: "...Made in Korea." May have had neck bow or accessories. As is $5-$10

GRAYHOUND ELECTRONICS INC. -5" seated white bear, pink ear linings and back paw pads, pink plastic nose, pink floss mouth. Pink velour shirt fastens in back with Velcro®. Shirt front says in black "Grayhound Crane Bear." Tush tag: "Grayhound, Specialty, Stuffed Toy, Made in Korea." Tag reverse: "Grayhound Electronics Inc, Toms River, N.J." *Courtesy of Shirley Taylor.*

GRAYHOUND ELECTRONICS INC. - 7" light brown and white koala, white snout, white belly and bottom, cream tail, plastic eyes, large oval plastic nose, brown felt paws. Ribbon tush tag: "GRAYHOUND, Specialty, Stuffed Toy, Made in Korea." Tag reverse: "Grayhound Electronics Inc, Toms River, NJ, Synthetic fiber." $5-$8.

HALLMARK CARDS INC (greeting cards/gifts) - A quartet of Hallmark bears. The 13" brown bear on left is not jointed but the limbs move freely. Side tag reads: "© 1984 Hallmark Cards, Inc., K.C., MO 64141, Made in Korea, 'Bea Bear'." Brown bear in the middle is 11", tag says © 1984, calls him "Bo-Bo Bear," Made in China, with original price of $8. The floppy white bear in the center is 7" from head to tail, has metal eyes and knitted scarf, tag says "© 1983, Made in Korea, 'Polar Bear'," original price $6.50. The 9" seated black bear on the right sports a red satin ribbon. In addition to his regular tag: "© 1990, Made in China," he also has a woven tag: "Designed Exclusively for Hallmark" and on the reverse: "By Heartline" in red. Although these bears are fairly plain, Hallmark now makes bears with more personality and accessories, like the one with an upside-down ice cream cone on his head (not shown here). Bears shown $6-$9 each.

HALLMARK GIFTS INC. - 7" white plush bear with tail, gray nose, plastic eyes, red neck bow. He holds a red cardboard box with white hearts. Paper ear tag: "Hallmark, Bearer of Gifts, © Hallmark Gifts Inc." Tush tag: "Designed Exclusively for Hallmark." Tag reverse: "Heartline, Manufactured for Hallmark ©1988 Heartline, Graphics International, Made in China." *Courtesy of Helen B. Evans, Billie's Emporium.*

HALLMARK - 7" white Hallmark bear with tail, red velvet neck bow, plastic eyes, gray velvet nose. Heartline side tag: "Designed Exclusively for Hallmark, © 1989 Graphics International, Made in Korea." *Courtesy of Rujean's Collectibles, Past and Present.*

HARD ROCK ® CAFÉ - 9" seated light brown plush bear with peach ear linings and paw pads, plastic eyes and nose, floss mouth. He wears a removable white shirt that says "Save the Planet, Hard Rock ® Café, Orlando." Leg tag: "Manufactured for Hard Rock ® Café Exclusively." Tag reverse: "Made in China." *Courtesy of Muriel Hoffman.* $10-$20.

HAMMS BEER - Hamms Beer Bear was used, circa 1970s-80s, in TV commercials. This black and white plush bear with red tongue is 17". "Hamms" is written on his chest. Body tag: "Dist. By Stephens Distributing Co., Minneapolis, MN. " Wire helps the bear stand tall; when you raise one arm the other arm raises too. *Courtesy of Ginny Kreitler.*

HARD ROCK®CAFÉ - On the left, 10" cinnamon brown plush bear with tan plush snout, back paw pads and ear linings, amber plastic eyes, black plastic nose with indented nostrils, black floss mouth, nickel sized tail, humped back. White cotton T-shirt reads: "Save the Planet, Hard Rock® Café, New York." Leg tag: "Manufactured for Hard Rock® Café Exclusively." Tag reverse: Year is unreadable, "Made in Korea." In the center: 5" all white kitten, light blue plastic eyes, pink plastic nose, 3 clear plastic whiskers on each side, red neck ribbon, holds red stuffed fabric heart. Round plastic button pinned to arm says:"No Drugs or Nuclear Weapons Allowed Inside, Hard Rock® Café." Tush tag: "PBC™// International// Div. of Pacific Balloon Co." Tag reverse: "PBC International-Ventura, CA....Made in China." On the right: 11" all white plush bear with plastic eyes, black felt nose, black floss mouth, straight back, black cotton knit T-shirt reads:"Hard Rock Café, Huntington Beach." The black plastic button pinned to shirt says: "Hard Rock Cafe//Save the Planet//We Recycle." Woven tag near tail: "Made Exclusively for //Hard Rock® Café" against an orange circle. Tag reverse: An "H" in a square double frame, "Harrington & Co, Inc.//© 1990 Beverly Hills, USA//...Handcrafted in China." 11" bear $12-$25, 10" bear $10-$20, 5" kitten $6-$10.

Harley-Davidson®

HARLEY-DAVIDSON® MOTOR CYCLES - 16" limited edition of 1,000 made by Cooperstown Bears. Dressed in pants, leather vest, and red bandana headband. Pictured in their 1998 catalog. A 1996 Harley-Davidson bear by Cooperstown Bears wore a hat, and a short-sleeved shirt under his leather vest. He was a limited edition of 1,000.

HARLEY-DAVIDSON ® MOTOR CYCLES - Front view of a trio of 7" Harley-Davidson "bean bag plush" bears made of brown velveteen with beige flannel snouts, back paw pads, and ear linings. Plastic eyes and nose, small tail. Paper and tush tags: "Official Licensed Product, © 1997 H-D, Produced by Cavanaugh Group International, Roswell, GA 30076, Made in China." On the left in red bandana and Harley T-shirt is "Roamer." His paper tag reads: "The freedom of the open road's What Roamer loves the best. He rides through mountains in the east, And deserts in the west." In the center wearing black boots and white/black Harley helmet is "Big Twin." Paper tag: "Big Twin's your friend, If you're ever in trouble, Just give him a holler, He's there on the double." On the right in black jacket with studs, is "Motorhead." Paper tag: "Everyone loves Motorhead, For riding he was born and bred, He cruises where the roads are wide, and shouts in passing 'Live to Ride!'" These were made for the 50th anniversary year only. Retail issue price $7.99 each.

Back view of 7" Harley-Davidson "bean bag plush" trio. Note the Harley-Davidson symbol on the back of "Motorhead's" jacket.

HARLEY-DAVIDSON ® MOTOR CYCLES - "Panhead Pete" the Leather Jacket Bear, 18" brown bear is ready for the road in leather pants, jacket, and hat. From the Harley-Davidson Plush Collection honoring Harley- Davidson Motor Cycles' 50th anniversary. Box has clear cellophane front and says: "© 1997, Manufactured by The Cavanaugh Group International." The third of the anniversary 18" bears was "Harley V Twin Bear." Further details unknown. *Courtesy of Trudy, Spencer Gifts, a Universal Studios Co.* Retail sales price $50 each.

HARLEY-DAVIDSON ® MOTOR CYCLES - "Cruiser Bear," 18" butterscotch plush bear wears orange headband, blue jeans, suspenders, and gray Harley-Davidson T-shirt. His box has a clear cellophane front. He was one of three boxed bears of this size, called the Harley-Davidson Plush Collection, made in 1997 to honor the company's 50th anniversary of the motor cycle division. Harley-Davidson Motor Cars was founded in 1903. Box: "© 1997, Manufactured by the Cavanaugh Group International." *Courtesy of Trudy, Spencer Gifts, a Universal Studios Co.* Retail sales price $50 each.

HENRY HEIDE INC. ("quality candies since 1869") - 12" brown plush, beige snout, open mouth, plastic eyes with black eyelashes, wears removable blue romper suit with "Heide" embroidered on a red diamond shape. Tag: "Heide Gummi Bear, 1988 Henry Heide, Inc., Russ Berrie & Co, Oakland, NJ." Original price $10.00. *Courtesy of Jean Laughery, Photo by Vivian L. Gery.*

H.J. HEINZ ™ COMPANY - 9" beige distressed plush, a limited edition reproduction of 1906 Teddy Roosevelt-inspired design. Heinz logo on white apron front reflects passage of the Pure Food Act, which laid the foundation for the modern food industry. Apron says: "Pure Food Products, 57 Heinz Varieties." Chef's hat has Heinz logo. Tag: "Heinz™ Bear, 1993, North American Bear Co., Chicago, Illinois." Available in 1994 for $25. *Courtesy of Jean Laughery, Photo by Vivian L. Gery.* Current price $150.

H.J. HEINZ ™ COMPANY - 15" beige distressed curl plush, suede paws, hand-embroidered nose, fully jointed, comes individually numbered and boxed. This ballpark vendor bear wears removable white pants, red/white striped shirt, red/white running shoes, red apron with "Pure Food Products, 57 Heinz Varieties." White hat says "Heinz." A wooden vendor box (which opens) hangs around his neck. He holds a plastic hot dog. The bear promotes Heinz ketchup. Released May 1998 by Douglas Cuddle Toys, Keene, NH. as part of their "Company Classics" collection. Photo was a pre-release magazine ad. Issue price $100.

Hershey's

For bear loving chocoholics, what could be better than bears advertising our favorite candies? Hershey's advertising bears are many and varied. And what other confection has two of its own theme parks? Hershey Park and Chocolate World, down the road from each other in Hershey, Pennsylvania are good sources for both chocolate and bears.

HERSHEY®'S - 7" dark brown bear with tan snout, ear linings, and paw pads, plastic eyes, floss nose and mouth, brown loop stitched to head. Body tag: "Ideal Toy Corp. 1982, Newark, NJ, Made in Korea." Note that he is larger than "Baby Bear Kisses." His head is more slender. He was not part of the three bear set, and was made in Korea. He is missing Hershey T-shirt and paper tag.. As is $7-$15.

5½" BABY BEAR "KISSES" - - 10" MAMA And 17" PAPA HERSHEY BEAR

Incredibly Priced - COMPLETE SET OF 3 BEARS . . . . $23.50 POSTPAID

These Are No Ordinary Bears. They Are Hershey Advertising Bears With Happy Smiling Vinyl Bear Faces . . . NATURALLY, THEY ARE MADE OF CHOCOLATE BROWN SOFT, SOFT PLUSH AND ARE SCENTED TO HAVE A DELICATE CHOCOLATE AROMA . . . . DRESSED IN COLORFUL T-SHIRTS . . . . Guaranteed To Make You Smile . . . . And We Feel That They Will Become An Important Collectible . . . DUE IN JUNE . . . ORDER NOW   MC/VISA ACCEPTED

TREASURE TROVE

19 VILLAGE RD. - MANHASSET, N.Y. 11030 PH# 516 627 4646

OUR 15th YEAR . . . . WE ARE THE *Source* FOR MODERN COLLECTOR DOLLS AND BEARS -
Please Send Long S.A.S.E. For Our Latest Illustrated Lists.

HERSHEY'®S (chocolate) - Copy of 1982 magazine advertisement for "Three Little Hershey Bears" by Ideal. The 17"plush Papa Hershey Bear and 10" Mama Hershey had vinyl mask faces. Their white shirts with Hershey logo were part of their bodies. Shirts for 5.5" Baby Bear Kisses came in white, yellow, red, and blue. *Courtesy of Kerra Davis.*

HERSHEY'®S - Ideal's 1982 "Baby Bear Kisses" is 5.5" seated, with an all fabric face, dressed in a removable yellow shirt. Shirt fastens with Velcro® in the back. His paper tag is shaped like a silver Kiss. Inside the tag it says: "Made of soft, safe, non-toxic plush fiber, surface washable. Sweet Scented with Chocolate Aroma - for sweet kids' sweet dreams!" and "Who's Sweet as a Kiss? Kisses, the Chocolate Bear!, The World's Most Unbearably Sweet Chocolate Cuddle, from Ideal." His ribbon tush tag: "© Ideal Toy Corp.1982, Newark, NJ, Made in Taiwan." Mama Hershey Bear, 10" seated, has a non-removable shirt, soft vinyl mask face, and the same ribbon tag as Baby's on her leg. Hershey's began making chocolate Kisses in 1907. $25-$35 each.

HERSHEY'®S - A quartet of sweet selections: 8" seated white plush bear in removable red "Hershey's krackel®" shirt, 18" standing beige firmly stuffed "carnival plush" bear in removable orange "Reese's™" shirt, 9"seated dark brown "Hershey'®s Milk Chocolate" bear with beige snout, and an individual 4" high silver fabric "Hershey's ™ Kisses" chocolate Kiss were all made by Acme in 1989. Krackel has red cotton fabric body under his shirt; Reese's body is orange fabric beneath his shirt. Reese's and Kiss were made in Korea. Krackel and Milk Chocolate were created in China. Krackel $15-$20; Reese's $15-$20, Milk Chocolate $10-$15, Kiss $3-$7.50.

HERSHEY'®S - 20" seated white "carnival plush" bear with brown nose, ear fronts, and back paw pads, plastic eyes, and floss mouth. The candy bar he's holding is a satin wrapper. Tush tag: "Acme™" and "© 1989 Hershey Foods Corporation, Trademark of Hershey Foods Corporation, Shell made in Korea." Paper tag on hand: candy bar replica on front and "Acme Premium Supply Corp., 4100 Forest Park, St. Louis, Missouri" on reverse. "Carnival" plush bears were discontinued in 1993/94. They got their name from having been prizes won at carnival booths. $10-$20

HERSHEY'®S - 9" seated brown plush bear with beige snout, back paw pads, and ear linings, black eyes and nose, floss mouth, no tail. This "Certified Hershey®'s Chocolate Lover" bear wears removable brown cotton overalls, engineer's brown hat, and red neckerchief with white dots. Foot tag: "Trademark of Hershey's Prod Corp., Charles Products Inc. Licensee, Rockville, Maryland 20852, Made in Korea." This little sweetie was available in 1986 and 1987 only in the Hershey's gift shop, not as a premium or by mail order. He sold for $9.95. Charles Products Inc. is no longer in business. $15-$20.

HERSHEY'®S - The 12" seated dark brown bear holds a 7" silver Kiss. His nose is flannel over a hard ball, eyes are brown plastic ovals, brown felt back paw pads. His tag reads: "KB" inside a diamond, "K.B. Bros. Inc, New York, NY, © 1990 Hershey Food Corporation, Made in China." The 8" chocolate brown bear holds a 4" silver Kiss. Leg tag has "KB" in a diamond and "Made in China." His triangle shaped face is a newer version. In 1998 the gift shop at Hershey's Chocolate World (celebrating its 25th anniversary) had KB bears holding green Kisses, as well as the usual silver or red. 12" $25 up, 8" $10-$15

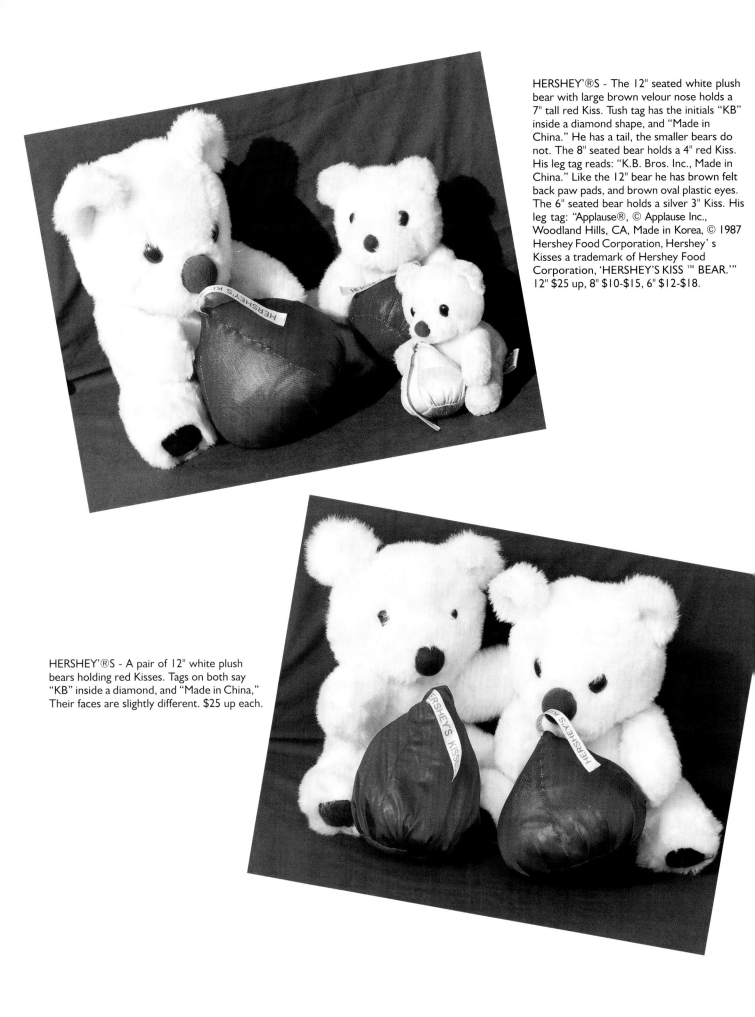

HERSHEY'®S - The 12" seated white plush bear with large brown velour nose holds a 7" tall red Kiss. Tush tag has the initials "KB" inside a diamond shape, and "Made in China." He has a tail, the smaller bears do not. The 8" seated bear holds a 4" red Kiss. His leg tag reads: "K.B. Bros. Inc., Made in China." Like the 12" bear he has brown felt back paw pads, and brown oval plastic eyes. The 6" seated bear holds a silver 3" Kiss. His leg tag: "Applause®, © Applause Inc., Woodland Hills, CA, Made in Korea, © 1987 Hershey Food Corporation, Hershey' s Kisses a trademark of Hershey Food Corporation, 'HERSHEY'S KISS ™ BEAR.'" 12" $25 up, 8" $10-$15, 6" $12-$18.

HERSHEY'®S - A pair of 12" white plush bears holding red Kisses. Tags on both say "KB" inside a diamond, and "Made in China," Their faces are slightly different. $25 up each.

HERSHEY'®S - Not only bunnies bring candy for Easter. This little brown Hershey's Kiss bear came along with a basket filled with candy, Easter 1998. *Photo courtesy of Tami Patzer.*

HERSHEY'®S - This little white Hershey's Kiss bear holds a red metallic fabric Kiss. She was a Teleflora Valentine gift accompanied by flowers in 1998. Her face is the newer version. (Other Hershey's Kiss gift ideas for Valentine's 1998 were Kiss shaped pendants in sterling silver or 14 karat gold, some with diamonds, by Hershey's licensee J & C Ferrara Co. Inc.) *Photo courtesy of Tami Patzer.*

HERSHEY'®S - This 8" seated white plush bear holds a silver Kiss striped with brown and its ribbon tag says "Hershey's Hugs ™" The leg tag has "KB" within a diamond shape and "Made in China." The paper Hershey's tag attached to his ear is a Hershey's bar facsimile on the front. On the back it says: "© 1994 Hershey Foods Corporation, Trademarks used under license, K.B. Bros. Inc Licensee." It gives the KB address as 200 Fifth Avenue, New York, NY 10010. One of the bear's eyes is silver and the other is brown. Paws have three floss toe separators. $15-$25

HERSHEY'®S - 15" standing brown plush bear dressed in removable brown shirt with blue neck and cuff trim, and "Hershey's" in white on the front. Tag: "Graphics International, KC, MO, ™ Hershey's Food Corp., Made in Korea." *Courtesy of Alison Hubbard-Miller, Photo by Bill Miller.*

HERSHEY'®S - 9" butterscotch colored handcrafted, unjointed, plush bear with humped back, chubby tummy, tiny tail, beige leatherette back paw pads, and pointed muzzle. He holds a molded plastic "Reese's Chocolate Peanut Butter Cup" with a big bite taken out of it. His molded red plastic tongue indicates that it's lip lickin' good. Vest and matching cap are black and white checked, with an orange Reese's patch on the vest left front. His leg spread is 10". Paper and tush tag information is the same as for Hershey's Chocolate Bear by The Blessed Companion Bear Company. Limited edition of 5,000. Issue price $29.95. One of three bears debuting the new bear company. (The third is a Campbell's Soup bear.)

HERSHEY'®S - 9" handcrafted, unjointed, chocolate brown plush bear with pointed muzzle, humped back, chubby tummy, leatherette back paw pads, and small tail. His leg spread is 10". Great attention has been paid to detail. He wears brown and white houndstooth cloth vest and matching cap. His red molded plastic tongue is licking his lips; as he holds a plastic replica of the Hershey'® Milk Chocolate Bar with chocolate squares protruding. His paper ear tag is also Hershey's bar shaped. Tag reverse: "…always offers a gentle reminder of the happiness of the simple pleasures of life." and "©1997 Hershey Foods Corporation, trademarks used under license, Name That Toon licensee, Manufactured by The Blessed Companion Bear Company™, San Rafael, CA., Made in China, Do not machine wash." Tush tag gives the same information. One of three bears crowning the company's debut into bear making. Limited edition of 5,000. Issue price $29.95. *Information courtesy of Craig Wolfe.*

Back view of 1997 "Reese's™ Milk Chocolate, 2 Peanut Butter Cups™" Bear, showing the patch on his back.

Back view of 1997 Hershey's Chocolate Bear by The Blessed Companion Bear Company showing the patch on his vest back - a black and white cow with "Hershey's Milk Chocolate," and "Made on the Farm."

HERSHEY'®S - On the left 12" tall "Homespun Hershey Kiss Bear" wears brown plaid dress with silk screened apron reading: "Hershey's" and showing three Kisses. On the right 15" distressed cream curl plush, 2nd limited edition (of 1,200 pieces) of "Homespun Hershey's Kiss Bear." She wears brown plaid dress, and a tea stained apron with "Hershey's" and Kisses machine embroidered on it. The basket comes with her. Both bears are fully jointed and wear plaid hair ribbons. Additionally the 15" has suede paws, a hand embroidered nose, and is individually numbered and boxed. Appeared in 1995 Douglas Cuddle Toys "Company Classics" catalog. A 5" miniature Miss Hershey Bear (not shown) was added in 1998. Issue price $16.50.

HERSHEY'™S - 11" tall, fully jointed, little "Hershey's Varsity Bear," wears a white jacket with black sleeves, black buttons, and a black "H." The black billed white cap has a silver Hershey's Kiss silk screened on it. Made by Douglas Cuddle Toys, © Douglas Company Inc., as part of their Company Classics series, 1995.

HERSHEY'S - 15" Hershey's Cocoa Bear seems to have come in from the cold and is warming up with a cup of hot Hershey's Cocoa. She is still half-dressed for the weather in her blue polar fleece jacket (red mittens attached) and red snow pants with embroidered logo. Ceramic mug screened with the Hershey's logo is included. 4th Limited edition of 1,500 pieces by the Douglas Company, 1998. Issue price $100.

HERSHEY'®S - Front and back views of 15" distressed brown curl plush "Hershey'®s Varsity Bear." He's fully jointed, has suede paws, hand embroidered nose. A limited edition of 2,400 pieces, he's individually numbered and boxed. He wears a lined, white wool jacket with suede-like sleeves and a matching cap. Embroidered logos on front and back of jacket and embroidered silver Hershey's Kiss on cap make him irresistibly sweet. By Douglas Cuddle Toys for their Company Classics series, 1995.

Left: HESS & BEMENT (candy, ice cream) - 13" printed cotton fabric cut-and-stuff type bear, firmly stuffed. Brown bear wears green and white striped jacket, "holds" printed ice cream cone in one hand and "Fowler's" candy box in the other. Hat says: "Hess & Bement." Shirt back says: "the Sugar Bear" in yellow. No tags. Date and place unknown. $8-$12.

Above: Back view of Hess & Bement cotton print bear.

HOLLAND AMERICAN (cruise ships) - 17.5" brown plush bear, wears removable white undershirt that reads: "Rotterdam" below a ship inside an oval. Paper tag: "S.S. Rotterdam." Bear purchased during the first UFDC (United Federation of Doll Clubs) sponsored "Doll Cruise '86." *Courtesy of Ginny Kreitler.*

HONEY BEAR FARM - Circa 1952, 9" tan cotton velvet body, bear wearing straw hat and removable blue/white striped overalls. Inverted triangle shaped head, large eyes, black front paws and shoes. Looks more like a pig than a bear. Paper sticker on front reads: "Visit Honey Bear Farm, Power's Lake, Genoa City, Wisconsin." *Courtesy of Ginny Kreitler.*

HUGGIES® (baby wipes) - 10" seated brown plush bear, pinkish brown plush snout, ear linings, and 3 paw pads. The fourth paw pad is white cotton with "Huggies ®" printed in blue. Tiny plastic eyes, nickel sized black velour nose, open mouth lined with tan flannel, no tail, slightly chubby tummy. Back tag: "DIAL-A-BEAR™, Warrington W., England, Made in China." (A customer in a thrift shop found Huggies first, but she said I needed him more, for the book, and let me buy him.) $15-$20

I

ICEE CORPORATION (frozen carbonated soft drink) - Circa 1977, 18.5" white plush bear, red shirt (with white "I" on front and "Icee" printed on the back) is part of the bear's body. ICEE bear trademark first used in 1965, and the first plush ICEE bear distributed in 1968. ICEE machines are generally located at convenience stores with a lighted ICEE bear display sign. Bears were available until 1991, redeemable with "points" printed on paper ICEE cups. *Courtesy of Ginny Kreitler.* Bear $24-$35.

ICEE CORP. - Photo shows a portion of a paper ICEE cup and the "proof of purchase" diamond - a point toward obtaining an ICEE premium.

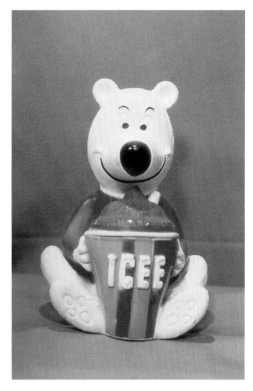

ICEE CORPORATION - Circa 1974, 8" white plastic figural bank of ICEE bear holding a cup of the "slurpy" "slushy" product. His red shirt is molded on. ICEE bank was free with enough "points" from paper serving cups. *Courtesy of Ginny Kreitler*. $25-$35.

INLAND - "Indy" bear, 23" tall, 18" arm span, 13" wide across his head. Brown and tan plush, dressed in denim overalls with gray denim tool apron sewn into it. Tool apron has embroidered label stitched into seam, reading "INLAND," followed by a circle, half of which is filled in with red. It is assumed that "Inland" stands for "Inland Steel." No tools in the apron pockets. *Courtesy of Alison Hubbard-Miller, photo by Bill Miller.*

INTERNATIONAL SILVER - 14" seated white nylon bear with tail, 14" leg spread, black velour eyes, black nylon nose, red floss mouth, red velour heart on each paw pad, white collar with red heart pattern and red ribbon bow. Bear holds (sewn to his hands) a red satin heart which says: (in white) "I'm sweet on you!" Tush tag: "International ® Silver Co." Tag reverse: "© 1993 All Rights Reserved." $5-$12

INTERNATIONAL SILVER - 7" seated brown bear with long pointed snout, small plastic eyes, black velour nose, red felt paw prints appliqued to back paws, red satin stuffed heart attached to neck ribbon says "Wild About You." Tush tag: "International ® Silver Co.." Reverse: "Made in China// © 1995 All Rights Reserved." May have been a gift with purchase of a certain silver item. $10-$16

INTERNATIONAL SILVER - 10" seated white nylon Santa bear, 7" across chest, red neck bow and Santa hat, red mouth, black eyes and nose. Tush tag: "International Silver Co., Made in China." *Courtesy of Helen B. Evans, owner Billie's Emporium.* $5-$12.

J

JORDACHE ® (jeans) - On the left 13" seated cinnamon brown bear with lambs-wool-like "fur," lighter tan snout and ear linings, amber plastic eyes, black plastic nose with indented nostrils. Cuffed jeans have round plastic disk simulating a belt buckle with raised horse head and "Jordache ® Country By Toyland" Two back pockets, one says Jordache ® and the other has a horse head stenciled-on in white. Leg tag is missing. On the right 11" white shaggy plush bear with tan snout and ear linings, plastic eyes, and small black plastic nose. Dressed in maroon velour vest and cuffed blue jeans. Pockets and belt buckle same as on 13". Leg tag: Sketch of horse head//"Jordache ®//Made in Israel by Toyland." Tag reverse: "© 1981 Toyland Div. of Caesarea Gienoit Ind. Ltd.// Jordache ® By Toyland//(A licensee of Jordache Enterprises.)" No hat . Each $12-$20.

Back view of Jordache bears showing their pockets.

JORDACHE/ LEE (jeans) - The 10.5" beige bear sold in 1983 at department stores. His removable belt and paper tag say "Jordache Country." Note his pink felt hat. The 14.5" white bear, marked "Lee" only on his jeans, came with several faces, grumpy, smiling. Paper tag shows bear in Santa hat with holly and red scarf with green fringe. A 1986 Christmas promotion at Glick's stores. *Courtesy of Ginny Kreitler.*

JOVAN (cologne) - 3.5" cinnamon plush, white snout and ear linings, plastic eyes with black felt surrounds, pink felt nose, mouth and head flower (with yellow felt center), white felt paws, peach flannel underside, pink cotton pouch sewn to body says: "Happiness is a Jovanimal." Undoubtedly this pouch held a small cologne purchase. Tush tag reads: "Especially For Jovan, Inc." Reverse: "Made in Korea." Jovan is a subsidiary of Coty US, New York, NY. Clear plastic tag holder in her ear may mean she originally had a paper ear tag. $4-$10.

K

KEEBLER ® COMPANY (cookies/crackers) - A close-up of 13" "Nicholas Bear" offered by Keebler Christmas 1994, for $9.95 and 2 UPCs from any Keebler products. The white plush bear wears a red Santa hat and a red shirt that declares in white "I believe in…." This bear is a replica of the bear in the new version of the movie "Miracle On 34th Street."

Copy of magazine advertisement that offered 1994 Keebler "Nicholas Bear."

KELLOGG'S - In 1925 Kellogg's offered a set of Goldilocks and the Three Bears as "cut and stuff" lithographed cloth dolls. The 12.5" Goldilocks has "Kellogg's" on her apron. The 12" Mama Bear holds a cereal bowl marked "Kellogg's" and 13" Papa Bear holds a box of Kellogg's cereal. They are from the 1925 set. A second set of similar cloth bears/dolls was made in 1926. The 9" Johnny Bear is from the 1926 set. *Courtesy of Ginny Kreitler.*

KELLOGG'S (cereals) - 10" and 9" seated all white plush bears, plastic eyes, black plastic nose, large red satin ribbon neck bow, head seams angled from ears to snout. Leg tag: "Kellogg's" in red script on white. Reverse tag: "Sasco Inc// New York, NY 10165// Promo Bear//™ Kellogg Company//....//Made in China." Bears have identical tags. $10-$18 each.

KENTUCKY FRIED CHICKEN - 16" Colonel Sanders bear, limited edition of 1,000, by Cooperstown Bears for their Americana Collection. White bear with eye glasses, goatee, black long ribbon bow tie, white jacket and pants, and holding a bucket with KFC and the Colonel's picture on it. Bear is pictured in Cooperstown Bears 1998 catalog. Harland Sanders opened his first restaurant in 1930, incorporated as "Kentucky Fried Chicken" in 1955, and died in 1980. PepsiCo purchased KFC in 1986.

KODAK (film) - A 1994 promotion of Kodak Royal Gold film offered this 9" brown plush "Photo Teddy Bear" with your favorite photograph printed on his blue-sleeved white T-shirt. Matching T-shirts or sweat shirts could be ordered for Teddy's friends. The offer expired December 31, 1994. Orders were directed to "The Smile Shop ®" No prices were given on the order blank as customers were advised to "check prices and service times at your retailer' s Photo Center." $5-$10.

KRAFT (cheese and other foods) - 13" standing, all white plush bear, plastic eyes, dark brown velour nose, brown floss mouth, white T-shirt with royal blue sleeves says (in blue) "Kraft" inside a red border and "Foodservice, ™," Tush tag: "A™, Acme ® Premium Supply Corp." Tag reverse: "Made in Korea." Bear came packed in a sealed clear cellophane bag. $10-$18

KRAFT - 18" brown plush pot belly "Toastem T. Bear" was offered by Kraft in 1981. His tan vest has a blue badge that says "I love Kraft marshmallows." A 9" hand puppet, in an identical outfit, was also available. *Courtesy of Ginny Kreitler.*

K-Mart

K-MART -17" white plush bear dressed for Christmas 1987 in red knitted hat and scarf. Tag: "© 1986 K-Mart Corp, Made in Taiwan " A bear shaped plastic hanging hook is sewn to his head. *Courtesy of 2nd Chance Home Furnishings.*

K-MART - 15" seated all white plush bear with plastic eyes and nose, wearing knitted cap with ear holes, a white pompom, and a reindeer, holly, and "Jingle Bear" knitted in. Tush tag: "© 1985 K-Mart Corporation, © Dan Dee Imports Inc., Jersey City, NJ." *Courtesy of Muriel Hoffman.* $10-$15.

K-MART (discount store) - 16" bear, from 1986, wears holiday red hat and scarf. The 15" bear in green hat and scarf is from 1987. It also came in a red outfit. The 1987 bears have the year knitted into the hat. Paper tags say "Our Christmas Bear." *Courtesy of Ginny Kreitler.*

K-MART - Paper tags say "150th Anniversary Charles Dickens' 'A Christmas Carol.'" The white plush couple (14" girl, 16" boy) is dressed in red, green, and plaid velvet with clothes, except for hats, as part of their bodies. Embroidered left foot dates them as "1993." Body tag: "Manufactured for and Marketed by K-Mart, Made in USA." *Courtesy of the Vivian Vanselow Collection.*

K-MART - Paper tag says "Santa's Magical Toyshop" on tag front, and "Santa's Magical Toyshop Bear Family bring their Christmas Magic to K-Mart for Christmas 1995" on the reverse. The white plush 16" elf bear with Santa hat and green velvet tool apron is ready to help. His clothes are part of his body. Left foot is machine embroidered in red "1995" Tag: "Designed Exclusively for K-Mart by Cheryl Ann." *Courtesy of Vivian Vanselow Collection.*

K-MART - Paper tag reads "A Teddy Bear Lane Christmas." The brown plush couple (both 20" standing) are clad in clothes which are part of their bodies. She's dressed to party in red and green velvet and lace. He's spiffy in his red and green corduroy outfit, red-banded derby hat, and pocket watch. The year "1994" is machine embroidered on their left feet. *Courtesy of the Vivian Vanselow Collection.*

L

L & K RESTAURANT - "Pierre the Bear" wears a red felt vest, green felt beret and scarf, and sports a jaunty black handlebar mustache. His personality definitely shows through. He was purchased in 1980 at an L&K Restaurant in Indiana. *Courtesy of Ginny Kreitler.*

LAND'S END (mail order catalog) - The "Eddie Bauer" bear, the first of a series. Serious looking goldish-brown bruin with amber plastic eyes, rust floss nose, and tail; dressed in green and navy horizontally striped shirt with white cotton collar. "Land's End" tag at shirt inside back. Tush tag: "Made exclusively for Lands End® Direct Merchants, by Gund ®" Tag reverse: "Gund Inc ® 1989, Edison, New Jersey, Made in Korea." Eddie Bauer began in Seattle as an outfitter for climbers and outdoorsmen. A Ford truck was later named for him. *Courtesy of Mary Duryea Kyviakidis.*

LAND'S END (mail order catalog) - 16" seated "Coach" teddy. His green rugby shirt is embroidered in navy letters "Coach." Workable whistle hangs around his neck. He's holding a matching figurine ornament of himself. Tag: "Made Exclusively for Land's End Direct Merchants by Gund, (1991) Limited Edition." Original price was $29.50. Various rugby team character bears from the fictitious "1926 World Champion Rugby Bears" team were produced for Land's End over a 5 year period. In 1993 "Coach" and "Big Daddy," the team owner, were both in the catalog. The rugby bears are no longer available. *Courtesy of Jean Laughery, Photo by Vivian L. Gery.*

LEE ® (jeans) - 15" seated cream plush bear wears a red bandana and blue denim jeans. A leather tab stitched to the waist band in back says "Lee." The paper tag in his ear shows a chubby bear bending over a sign board that says "Lee Teddy Bear" and "Lee." In 1985 he was purchased at a department store for $4.98. Current price, with all tags intact, $15-$25.

 Left: ® M.R. LEE (jeans) - 14" standing chocolate brown plush bear, tan snout, ear linings, and back foot pads, brown velour nose, brown floss mouth, amber plastic eyes. Heavyweight removable red sweatshirt with navy blue neckline and sleeve trim, "® M.R.Lee" machine embroidered on shirt front outlined by a white frame. Shirt closes at neckline in back with Velcro® strip. Heavyweight removable dark blue denim jeans with belt loops (no belt) and side pockets, two open back pockets, right pocket says "Lee ® M.R." and leather square above it at belt line says the same. Ribbon tag at back waist seam: "® M.R. Lee" Tag reverse: "ACI International, Made in China." *Courtesy of Carol and Edith Rinehart Ford.* $15-$25.

Above: Back view of Lee jeans bear.

LEE ® (jeans) - 15" light beige plush "Lois Lee" bear is unjointed and has no identifying tags. "Lee" is stitched on the chest pocket of her Lee denim overalls. She was a giveaway in 1989 with a purchase of Lee jeans.

LEVER BROS., SNUGGLE ® - 11" Snuggle with suction cups on his paws, and wearing a red-sleeved white T-shirt that says "Go Big Red" on the front. Regular "1986, Snuggle, Lever Bros, Russ" tush tag. Shirt has "Russ Berrie" tag sewn into inside seam. $20

Lever Brothers' "Snuggle"

LEVER BROTHERS CO., SNUGGLE ® FABRIC SOFTENER - 7" tan plush bear, stiff, with amber plastic eyes, small black plastic nose, emphasized muzzle with open black mouth showing red felt tongue. This was the first Snuggle issued in the US, in 1983. Leg tag: "Snuggle ®, Made only for Lever Bros. Co." Back of the tag is undated, but identifies Russ, Berrie as the maker. On the left is a 3.5" Christmas ornament Snuggle with plastic eyes, muzzle, and hands. Date unknown. In the center is "Safety Book From Snuggle, starring Snuggle the Bear" © 1983, 1989 Lever Brothers Co." It's a story-activity book on how to react in safety situations. Book $5, 7" $8-$12, 3.5" $15-$25.

LEVER BROS., SNUGGLE ® - 20" cream Snuggle backpack with white shoulder straps and a zipper down his back. Amber plastic eyes, black plastic nose, black felt mouth with long red felt tongue. Ears and velour muzzle are light tan. Back seam tag: "Snuggle ®, © 1982-85 Lever Brothers Company, Made Exclusively for Lever Brothers Company." Tag reverse: "Trudy, Norwalk, CT, Sewn in Haiti." $20.

LEVER BROS., SNUGGLE ® - Second issue Snuggle. 10.5" lamb-like fur, 25-cent piece size plastic ear tag: "Russ." Paper ear tag is dark and light blue, bordered in hot pink, with the name "Snuggle" and Snuggle's head. Tag reverse: "Russ, Snuggle © 1986 Lever Brothers Company, Made in Korea." Tush tag: © 1985." (Notice earlier date.) Original CVS price tag $12.00. In 1986 the bear could also be obtained for $9.95 and 2 UPCs. It arrived with a thank you note "for using Snuggle fabric softener and for participating in the 'Snuggle' offer." In the "Surf & Snuggle Bear offer" in Sunday supplement ads of 1986, you could get a 14" Snuggle Bear for $4.95, a UPC from Snuggle Fabric Softener and a 64 ounce bottle of Surf, plus your cash register receipt. Snuggle appeared in a "Parents" magazine ad for Healthtex in 1996, and a 1997 ad with the "Official laundry products of Little League Baseball." *Some information courtesy of Jean Laughery, Bear courtesy of Shirley Taylor.*

LEVER BROS., SNUGGLE ® - 16" Snuggle in a special promotion shirt. Plastic "Russ" ear tag. Blue-sleeved shirt says "I (heart) New ultra soft Snuggle ®" The same wording is used on shirt front and back. On shirt front "Snuggle" is written in blue, and on the back in green. Bear came packaged in clear plastic bag. Tag on the back of his shirt reads: "Tiny Tees ®, Made in China, Shirts Illustrated Inc, Santa Barbara, CA." Ads for "new ultra soft Snuggle" appeared in 1989. $25.

Back view of Snuggle wearing shirt promoting "New ultra soft Snuggle ®."

LIMITED COMMODITIES INC - 15" cinnamon brown bear, nose same color, plastic eyes, tan ear linings and back paw pads, red neck bow (probably not original). Tush tag: "© 1988 Ltd. Commodities, Inc., "Preppy Bear"™, Made in Korea." With his name, he may originally have been dressed in a "preppy" outfit. *Courtesy of 2ⁿᵈ Time Around.*

LEVER BROS., SNUGGLE ® - 14" Snuggle Bear Puppet was offered for $7.95 with 2 proofs-of-purchase from any Snuggle product, and the original cash register receipt.. Suggested retail price $20. Offer began March 6, 1994 and expired June 30, 1994 "or while supplies last." Limited to one order per family. The item being promoted was "Snuggle Singles," individual sheets of fabric softener. Tush tag says: "© 1993 Lever Brothers Company, All rights reserved." Tag reverse: "Greystone International, Shelton, CT, Made in Korea." Note the paper tag is different from the tag on the 10.5" bear. This tag has a wider border, and the bear and word "Snuggle" are larger. The reverse of the paper tag also says "© 1993."

LEVI STRAUSS & COMPANY (jeans) - The durable denim jeans concept was conceived during the California Gold Rush of 1849 by Conestoga wagon-top maker Levi Strauss and hasn't waned since. This 12.5" beige bear wears Levi Overalls and a red/white polka dot neckerchief. Red "Levi" tag at left waist. A great flea market find. *Courtesy of Ginny Kreitler.*

LIPTON ® TEA - 15" golden bear, dressed in a full-length tea print dress with red and white plaid trim, and matching hair-bow. Her embroidered bodice and Lipton Tea cup let everyone know her preferred beverage. She has suede paws and hand embroidered nose, is individually numbered and boxed. Limited edition of 1,200 by Douglas Cuddle Toys, Keene, N.H. in their Company Classics series, 1995.

LOVABLE BRAS/ HERSHEY'S - On the right 8" thumb-sucker bear called "Baby Bear" has a vinyl mask face. He was available in 1976 with the purchase of a "Lovable BARE Bra" (lacy and see-through sheer) for $2.50 and the coupon from the bra package. Many of these little bears are unmarked or missing tags. On the left is a 5" Hershey's Kisses bear from 1982 in a removable red T-shirt, with his paper Kisses shaped tag. Other shirt colors available for him were white, yellow, and blue. The original "Kisses" bear was part of a set of three with Mama and Papa Hershey. They were scented with the aroma of chocolate. *Courtesy of Ginny Kreitler.*

M

McCRORY CORP. (5&10 cent stores) - In center 15" seated red plush bear with white snout, ear linings, and back paw pads, plastic eyes, red velour nose. Leg tag: "Made in Taiwan Expressly For McCrory Corp., York, PA 17402." On the right 4" pale yellow bear with white snout and ear linings, tiny plastic eyes, pink plastic nose, paws sewn in prayer position, lavender and white striped night cap, matching nightshirt is part of bear's body. Tag , nearly as big as the bear, says: "Made in China Expressly For McCrory Corp, York, PA 17402, © YDC." 15" bear $5-$8, 4" bear $2-$6.

McDONALD'S ® (fast food restaurant) - (a) 4" standing black and white flannel panda, brown floss eyes and toe marks, no tail,. Leg tag: "M" (golden arches) with "McDonald's ®" across it. "Made in China" below. Tag reverse: "© 1997 McDonald's Corp., Made in China." (b) Brown bear on all four feet with short tail, black floss nose and eyes, tag same as panda. Two in a series of many endangered species animals in a "save our wildlife" promotion. $2 to $6 each.

McDONALD'S ®- 7" seated gold curly plush, plastic eyes with pink eyelids, ball-like pink felt nose and pink felt tongue sewn to red felt open mouth, pink felt ear linings, wears red felt brimmed hat with 2 green felt holly leaves and ear holes, and green flannel neck scarf with red yarn fringe. Back waist tag: "© 1987, Henson Associates, Inc.,Baby Fozzie Bear is a Trademark of Henson Associates, Inc., All Rights Reserved." Tag reverse: "Simon Marketing Inc., Los Angeles, CA, Made in China." Paper tag: "McDonald's ® Presents Jim Henson's Baby Fozzie Bear™ © 1988 McDonalds". Baby Fozzie is one of the Muppets (puppets). *Courtesy of Peggy Monahan.* $15-$20.

J.E. MAMIYE & SONS - 9" seated black and white plush panda, plastic eyes, velour nose, eye surrounds are black plush, black tail, red ribbon neck bow. Tush tag: "J.E. Mamiye & Sons, New York, NY 10016." Tag reverse: "Made in China." Type of company and date are unknown. Well made. Courtesy of Muriel Hoffman. $8-$15.

MARS, INC. (candy bars) - 17" chocolate plush bear, beige snout, ear linings, and back foot pads, amber plastic eyes, chocolate flannel nose, silver dollar sized tail, wears white sweatshirt with green neckline, sleeve and bottom trim, says "Milky Way ® Bar" in green machine embroidery across shirt front. Leg tag: "© Mars, Inc. 1987//Graphics Int'l Inc.//Made in Korea, S7502." The S number may be an order number. $15-$25.

MARS, INC. - Tan plush bear with cream muzzle, ear linings, and back paw pads, plastic eyes and nose. He wears a yellow shirt with "M&M" in brown on it .This is the bear, given by a 5 year old to his grand-mother Frances Pew Hayes in 1984, that began the collection which gave birth to the Teddy Bear Museum of Naples, Florida. The Museum's current director is one of Ms. Hayes 7 children. *Courtesy of George B. Black, Jr., Teddy Bear Museum .*

MARS, INC.- 14.5" chocolate plush bear, flesh colored nose, wears shirt with "M&M" letters in white. He has a chubby tummy and large feet. Tag: "© Mars, Inc., 1987, Graphics Int'l Inc, K.C., Mo 64141, Made in Korea, #S 7500." During World War II, M&Ms ("the chocolate that melts in your mouth, not in your hands") were a C-ration dessert. They have traveled aboard NASA's space shuttles, and served as a sponsor for the 1984 Olympic Games. *Photo courtesy of Ann Christensen Denney*

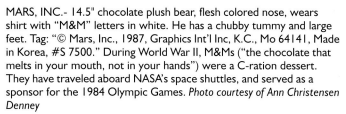

MARS, INC. - Pair of 2.5" molded and flocked figures of the "Snickers" Bear and the "M&M" Bear, in applied outfits identical to their larger stuffed bear versions. Incised on bottom of foot: "Made in China, © Mars Inc. 1987." *Courtesy of Oneida Callaway, Photo by Mary Ann Callaway Dennis.*

MARS, INC. - 16" butterscotch plush bear with cream snout, ear linings, and back foot pads, amber plastic eyes, rust velour nose, wears rust sweatshirt with royal blue neckline, sleeve and bottom trim, white patch with blue "SNICKERS ©" on white and "Bar" in blue on shirt. Leg tag: "© Mars Inc 1987// Graphics Int'l// Made in Korea, S7501." $15-$25.

MARSHALL FIELDS/ HALLMARK - 15.5" white bear on left is from Marshall Fields department store in Chicago, purchased in 1986. Body tag: "Marshall Fields", white knit stocking cap has ear holes and "Mistletoe" knitted across the front. On the right is 17" (sitting) Hallmark bear. He wears red knit sweater with white snowflake design. Purchased in Hallmark card shop in 1986. *Courtesy of Ginny Kreitler.*

MARSHALL FIELDS (department store) - 12" cream plush bear dressed in running suit with striped leg warmers and maroon head band. Maroon shirt says "Marshall Fields" in white letters. Paper ear tag: "R.Dakin." *Courtesy of Oneida Callaway, Photo by Mary Ann Callaway Dennis.*

MAX FACTOR (cosmetics) - This 8" white bear (similar to Coca Cola polar bear) came with 3 shades of Max Factor eye shadow at Christmas 1986. The removable red tank top shirt says "I (heart) MAXI" in white. *Courtesy of Ginny Kreitler.*

MAXWELL HOUSE COFFEE - In 1971 Maxwell House Coffee offered 3 sizes of brown plush bears for the total price of $4.95. The 2 bears shown here are 21" and 13". The red and white striped waist length nightshirt is part of the body. The striped nightcap with red pompon is sewn on. The coffee got its name from the Maxwell House Hotel in Nashville. The coffee's slogan "good to the last drop," is attributed to a comment made by Teddy Roosevelt. *Courtesy of Ginny Kreitler.*

MERCURY LUGGAGE MFG CO. - 7" flat aqua stuffed bear, white snout, plastic eyes, black plastic nose, arms and legs positioned sideways. Back seam tag: "Mercury Luggage Mfg. Co., acrylic fiber, Made in Taiwan." No date or location known. $4-$8.

MERVYN'S (department stores) - 20" bright yellow plush bear with sheared snout, plastic eyes, black floss nose and mouth, wears yellow knitted sweater with yellow hearts on black sleeve and chest stripes. Back waist tag: "Made Expressly For Mervyn's, Made in Korea." Tag reverse: "Synthetic fibers." Softly stuffed and very huggable. $8-$18.

MEMORIAL WREATH COMPANY - 6" black and white plush panda, flesh colored plastic snout with freckles and painted nose, plastic eyes with black felt surrounds, flesh colored plastic hands with thumbs extended (for sucking but do not fit into mouth) and bare feet, nickel-sized white tail. Bottom feels weighted. Tush tag: "Memorial Wreath Company." Tag reverse: "Synthetic Fibers & Ground Nut Shells, Made in Korea." Date and place of company are not given. $3-$8.

THE MEDICINE SHOPPE ® (drug store) - 9" seated light brown bear with cream snout, ear linings, and back paw pads, plastic eyes, black plastic nose, black floss mouth. Humped back makes his head protrude forward. White polyester and cotton T-shirt has "The Medicine Shoppe ®" across the front in blue. Tush tag: A crown shape and word "Crown." Tag reverse: "Made in Taiwan, R.O.C." $5-$8.

MICHAELS STORES INC- 3,5" tan jointed bear, plastic eyes, floss nose, mouth and paw claws, red ribbon neck bow, gold hanging cord on head, wearing straw hat with metallic gold tag: "Made in Philippines." Paper tag: "3.5"// Twisted// Teddy Bear// 977009// Michael's ® Stores Inc.// Made in China." Original sales price $1.29. Current $2-$7

MICHAELS (a hobby store) - 11" light beige plush bear, wears maroon wool scarf with "Michaels" in gold script, has plastic eyes and nose. The bear was purchased on-site at Christmastime 1987. *Courtesy of Ginny Kreitler.*

MIDWESTERN HOME PRODUCTS - 6" tan bear with white snout and ear linings, plastic eyes and black plastic nose, floss mouth, stiff body, raised arms, arm span 4.5". Back tag: "Distributed by// Midwestern Home Products// Wilmington, Delaware 19803." Reverse: "Made in China." 6" green Christmas print fabric rocking horse's tag reads the same. Purchased in Indiana. Horse $3-$6, Bear $3-$6, Set $7-$10

MISS BOUTIQUE/ McCRORY CORP. - (#1) 11" milk chocolate brown plush bear with tail, molded plastic face and "people" hands with extended thumbs which fit into mouth, black painted eyes and nose, brown freckles, pink lipped open mouth, painted brown arrow shaped eyebrows, flattop head. Leg tag: "Miss Boutique, NYC, Made in Taiwan." (#2) 11" dark brown plush bear, very like bear (#1) but has rounded top of head, narrower forehead, and a more slender face and body. Back seam tag: "Made in Korea Expressly for McCrory Corporation, York, PA17402." (In 1977 Russ Berrie made a similar 5" dark brown bear (not pictured here); tag read: "Bippi, © Russ Berrie, synthetic fibers, Made in Korea.") These little thumb-suckers are similar to the Lovable Bare Bra bears. $5-$12 each.

MODE O' DAY (department stores) - 13" beige plush bear, pointed white snout, amber plastic eyes, black plastic nose, wears red cotton shirt with "MODE O' DAY" in white on the front, and "Happy Holidays" on the back. Leg tag: "Smith & Margol." Tag reverse: "Made in Taiwan." *Courtesy of Lavonne Morrell*. $10-$15.

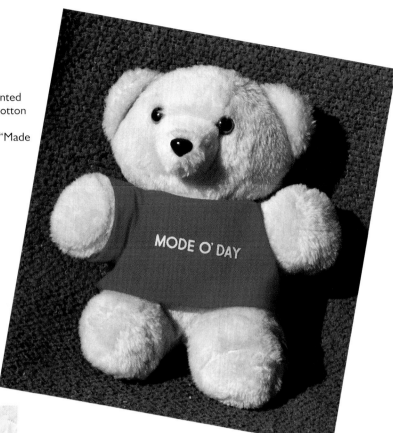

Copy of original 1997 newspaper ad for limited edition "Montgomery" and his little "bean bag bear" brother.

MONTGOMERY WARD (department store) - 13" dark brown plush bear with beige snout, ear linings, and paw pads, plastic eyes, brown velveteen nose, wears candy-cane striped nightcap with white ball tassel and "Montgomery" machine embroidered in script on white cuff of hat. Back seam tag: "Montgomery ™" Tag reverse: "© 1997 Commonwealth Toy and Nov Co Inc, NY, Made Exclusively For Montgomery Ward, Made in China, Printed in Hong Kong." Paper tag (3.5"x4.5" folder): "Montgomery," color picture of Montgomery bear and "1997." Inside paper tag: "This comes with hugs and kisses, and lots of love to say…Sure do hope that you'll have A Very Special Holiday! Your Buddy (paw print), Montgomery ™." Tag back: "Made Exclusively for Montgomery Ward, Limited Edition 1997, Made and Printed in China." Original price $12.99. A small size bean-bag Montgomery was also made and sold for $4.99.

MORTON ® SALT - "When It Rains Bear" is a 15" fully jointed white distressed curl plush girl bear, with suede paws, hand-embroidered nose. Limited edition of 2,400, individually numbered and boxed. She wears an old-fashioned navy woven plaid dress with muslin trim. The Morton® Salt girl and umbrella appear on her dress yoke and the words "When it rains, it pours" is in blue script on a white strip around the skirt just above the hemline. Bear comes with a white muslin white-handled umbrella which says "Morton Salt" below a girl with an umbrella. The 11" white bear is also fully jointed. Her dress, except for the yoke, matches the 15" Morton bear but the little bear wears an apron showing a silk-screened girl with an umbrella and "when it rains, it pours" in blue script lettering. Both sizes wear a blue ear bow. Made by Douglas Cuddle Toys, 1995.

Above: THOMAS D. MURPHY CO. - 8"seated cinnamon brown plush bear with tilted head, tan snout, ear linings, and back paw pads, black floss mouth and toes; tail, rounded back. Royal blue T-shirt has: "The Thos. D. Murphy Co., Red Oak, Iowa" printed in white on the front, and "100, One Hundredth Year Anniversary 1889-1989" on the shirt back. Leg tag has crown shape and word "Crown, ASI 47679" Tag reverse: "Made in Korea." $5-$10

Left: Back view of Thomas D. Murphy Company's 100th anniversary bear.

NABISCO BRANDS (candy, cereal, cookies) - 17" standing cinnamon brown bear, soft and cuddly, blue plastic eyes, chocolate brown velour nose, silver dollar sized tail, removable orange sweatshirt with yellow ribbed neckline and bottom, says "BUTTERFINGER ®" in bright blue on shirt front. Leg tag: "© 1987 Nabisco Brands, Inc.//Graphics Int'l Inc//Made in Korea." Plastic ear tag: "Heartline because you're special" with red heart and red rainbow shape. Paper ear tag: "Life's a whole lot sweeter with Chocolate Chums" and inside: "I'll never let a chance to be with you slip through my fingers! Confectionately yours, CRUNCHLES" Having all its tags and marked shirt increases its value. *Courtesy of Edith and Carol Ford.* $20-$30.

NABISCO BRANDS CO. - 5" seated cream bear, tiny plastic eyes, small black felt nose, red neck bow, round ship's lifesaver motif printed on paper tag around neck, which says in red: "You're My Lifesaver." Bear came inside a 5.5" round metal can with red, orange, green, yellow, red stripes and "LifeSavers, Assorted" on two sides. Paper sticker on the can lid reads: "This collector's tin also contains an assortment of individually wrapped LIFESAVERS, Assorted, Net Wt. 4.4 oz." Paper label on can bottom: "Lifesavers ® is a trademark of Nabisco Brands Company. Used under license © 1995 by Nabisco Brands Company. Distributed by American Specialty Confections, Inc., Lancaster, PA 17603" Bear was purchased in 1997 from Figi's mail order catalog for $19 (including shipping).

NABISCO BRANDS INC. - 15" cinnamon brown plush bear, velvet snout, ear linings, and back paw pads, brown plastic nose, indented mouth, blue plastic eyes, has silver dollar sized tail. Leg tag: "© 1987 Nabisco Brands Inc, Graphics Int'l, Inc., Made in Korea." Huggable, floppy, with chubby tummy. May have worn a T-shirt advertising a Nabisco product. He's similar to the "Butterfinger" bear. *Courtesy of Peggy Monahan.* $7-$10 as is.

NATIONAL VIDEO ™ - 10" seated dark brown bear, tan snout, ear linings, back paw pads, plastic eyes, black plastic nose with indented nostrils, floss mouth, white T-shirt with red sleeves and "National Video" printed in green and red on shirt front. Shirt tag: "Made in Korea." Leg tag: "B J Toy Co., Inc.// Pen Argyl, PA.// Shell Made in Korea." Rickrack hanging loop sewed into back seam at neck. Center front and back body seams, seams from ears to snout form a triangle. Arm and leg spread 9". Undated. $5-$8

NATIONAL SUPERMARKETS - Items deemed a good value are called "The Presidents Choice, and these two (plus Mamma) were his choice bears in 1986. Father Bear appears on many National brand children' s products like diapers, cookies, cereal and such. Father, 16" sitting, wears a removable square-necked sweater with "Teddy" in black. Golden brown "T.J. Bear" is 8" tall. Bodies are tagged on the left hip "National Products, Presidents Choice." Shown with "Swedish Ginger Teddy Bear Cookies." *Courtesy Ginny Kreitler.*

National Supermarkets' 12" (sitting) Mamma Bear of 1986 in a yellow sweater with lace collar and sleeve edge. "Terri" is spelled out in black across her sweater front. *Courtesy of Ginny Kreitler.*

THE NEWS-SUN - 19" standing brown plush bear with tan belly, snout, ear linings, and back paw pads, black velour nose, black flannel sewn mouth, 1" oval plastic eyes, flat 2" tail. Wears removable royal blue cotton T-shirt with red setting sun and "Sunshine" and "The News-Sun" in gold lettering. Leg tag: "Art's Toy Mfg Co.Inc, Made in Korea" with bear logo. Tag reverse: "© 1988 Art's Toy Mfg Co Inc, 673N 13th St., Easton, PA" (Purchased in Illinois) $8-$15.

NIKON ® (camera) - 6" beige plush bear, white snout, ear linings, and back paw pads, plastic eyes, black velour nose, black floss mouth, wears white T-shirt with yellow sleeves. "Nikon ®, We take the world's greatest pictures ®" is printed on the shirt front. Leg tag has bear logo with "Greek" on his shirt and "ASI 62960" Tag reverse: "Made in Korea." Undated. $6-$10.

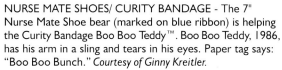

NURSE MATE SHOES/ CURITY BANDAGE - The 7" Nurse Mate Shoe bear (marked on blue ribbon) is helping the Curity Bandage Boo Boo Teddy™. Boo Boo Teddy, 1986, has his arm in a sling and tears in his eyes. Paper tag says: "Boo Boo Bunch." *Courtesy of Ginny Kreitler.*

NORTH AMERICAN VAN LINES/ COOK COUNTY SAVINGS/ KADI RADIO - Here's an interesting trio of bruins. On the left 8" dark brown bear, from 1976, with yellow snout and eyes looking up. His white shirt bears the logo "North American Van Lines" in blue below a red and blue ball with a left pointing arrow. In the center is a worried looking 19.5"bear with "Cook County Savings" on his foot tag. (Worried is not a good look for a Savings & Loan guy.) Note the black pompom navel. The removable red-sleeved white shirt of the bear on the right reads "KADI Radio 96, Bear This in Mind." *Courtesy of Ginny Kreitler.* Bears $15-$20 each.

ORIENTAL TRADING CO. INC. (mail order catalog) - 9" seated tan bear dressed as drum major in red flannel jacket and blue plush pants which are both part of his body. Black stuffed hat with yellow felt plume, yellow yarn shoulder fringe, black belt with yellow felt buckle. Oval plastic eyes, round black nose, black floss eyebrows and smile. Leg tag: "Oriental Trading Co. Inc., Omaha, Nebraska 68137, Made in China." $5-$10

OWENS CORNING - This 16.5" "Big Pink Bear" was found at a home show in 1979. "Big Pink My Home Is Insulated With Pink Owens Corning Fiberglass" is printed on the bear. Navy blue overalls are part of the body, which is made from a fleece-like material. Ivory snout and paw pads, plastic eyes, and pink nose add to his appeal. *Courtesy of Ginny Kreitler.*

OSH KOSH B'GOSH ® (children's jeans/ toys) - Pair of 10" standing terry cloth bears. She is cream with brown features, a red/ white striped body, and OshKosh B'gosh ® denim jumper with yellow buttons on the straps. Her part-of-her-body beige shoes tie with floss shoelaces. He is light brown, with black floss features, blue/white striped body, floss-laced brown sewn-on shoes, and denim OshKosh B'gosh ® overalls with red buttons on the straps. Each has a red "OshKosh B'gosh®" patch on the bib. Back of neck tag: "OshKosh B'gosh ®" Tag reverse: "© Eden LLC, NY, NY, USA, Made in China." Paper arm tag on each: "OshKosh B'gosh ®, The Genuine Article." Tag reverse: "Authorized Representative Eden International Limited, 180 Strand, London, England, © Eden, New York, NY, Baby B'gosh ® and OshKosh B'gosh® are registered trademarks of OshKosh B'Gosh Inc., Oshkosh, WI, Made in China." (Eden also is the US distributor for Paddington bears.) $10-$15 each.

PAMPERS (disposable diapers) - This 13" light blue cotton polyester velour bear has a ribbon around his neck with the Pampers bear printed on it. Back tag: "Lemon Bear Inc. 1985." *Courtesy of Ginny Kreitler.*

PAMPERS/ ISALYS CO. - Pampers offered a white long-haired Anniversary Bear, on the left, which came holding a figure 1, 2 or 3 in his paws. Paper tag: "Animal Fair." The "Klondike Hugging Bears" on the right were offered by Isalys Company in 1985 and 1986 for $6.45 (including shipping) and 3 UPCs from 6-pack packages of Klondike Ice Cream bars. Velcro® on 10" white plush Mother polar bear keeps 5.5" Baby bear attached. Tag: "Sun Toy Corporation, Long Beach, CA 90805." *Klondike promotional information courtesy of Jean Laughery. Photo Courtesy of Ginny Kreitler.*

PARCO FOODS INC - 12" tan plush bear wearing green felt vest and yellow polka-dot bow tie. This teddy was attached to a lithographed tin when purchased in 1987 from Wanamaker's Department Store. The tin of "Sweetie Bear Cookies" with bear cost $10.00. Tag: (bear logo) "TRUDY, Norwalk, CT USA." *Courtesy of Jean Laughery, Photo by Vivian L. Gery.*

J.C. PENNEY (department store) - 24" tan plush bear with golden brown velour snout and paw pads, dark brown plush nose, nickel-sized plastic eyes, floss mouth, plastic hanger sewn to head, red velour mittens on red cord hanging around his neck. Tush tag: "The J.C. Penney Collection// J.C. Penney Co, Inc.// Dallas, TX 75265-9000" Tag reverse: In English and Spanish, materials and "Made in China." A Christmas bear, year unknown. $5-$15.

PENCO FINANCE - 11" standing light brown plush bear with cream snout and back paw pads, plastic eyes and black plastic nose, black floss mouth. Wears white T-shirt with navy blue sleeves and neckline. Patch on shirt front says in blue "PENCO Finance Company." Leg tag: "TRUDY, Norwalk, CT USA, Made in Korea, © 1983 Trudy Toys." (The basic bear looks like the one used for State Farm Insurance bears.) $5-$8

J.C. PENNEY - Christmas Bears 1987. The 16" standing tan bear on the left says "J.C. Bear" on his removable green and white sweater. The 24" rust-colored bear on the right is called the "Bear With No Name." Back tag: "Made Exclusively for J.C. Penney by Gund." Tag reverse: "Gund ®, Gund Inc. © 1987, Made in Korea" Left ear tag: "Gund" & reverse: "JC Penney." Paper tag reads: "Take Me Home and Love Me and Give Me a Name." Wears black ribbon tie. The bear's price was $40, or $15 with a $50 purchase. *Information courtesy of Jean Laughery, Photo courtesy of Ginny Kreitler.* Current price $20-$25 with all tags intact.

PEPPERIDGE FARM™, INC. - 17", fully jointed, beige distressed curl plush distinguishes the Pepperidge Farm Bread Bear. She wears glasses, a white bibbed apron, that says "Pepperidge Farm," over her gold top and autumn toned patchwork skirt. Introduced in 1998 by Douglas Cuddle Toys as part of their Company Classics series. Ensemble comes with bear, table, cloth, bread sculpture, tools, and kitten. Issue price $130.

PILLSBURY - 13" standing white plush bear with blue eyes, nose, and mouth, wears a bib with a Pillsbury doughboy, doughgirl, and the name "Cupcake" printed on it. Body tag says "1974." *Courtesy of Ginny Kreitler.*

PILGRIMS PRIDE CHICKEN PRODUCTS/ SPORTO (sports clothes) - On the left 16" light brown plush "Tender Bear" in his red/white/blue jogging togs with his name on his headband was offered by Pilgrims Pride Chicken Products in 1986 and 1987. On the right 13" "Sporto" bear shows off his red stretch acrylic body and colorful clothes: red/lavender/white striped sleeveless shirt, knitted purple leg warmers, and purple tank-top which says "Sporto" in white. Purchased in 1984. *Courtesy of Ginny Kreitler.*

PINE SOL (liquid household cleaner) - The 15" bear on the right, made of a long brown fur fabric, sold for $9.99 in 1978. Bottom tag: "Mighty Star Ltd, Montreal, Que, Made in Canada." Paper tag on ear: "Exclusively for Pine Sol," is shaped like a pine tree, yellow with a green frame. On the left is the 12" standing beige bear. Tag: "Trudy, 1983." The bear was offered again in 1986 and named "Benjamin." *Courtesy of Ginny Kreitler.*

PRECIOUS MOMENTS (porcelain figures of children, and greeting cards) - 11" grayish tan plush bear with white snout and back paw pads, pink ear linings and pink satin toes, brown plastic nose, winsome green and white large plastic tear drop shaped eyes, 2.25" mounded round tail, oversized pink satin polka-dot tie. Metallic gold cord around neck suspends pink plastic heart locket with "Precious Moments" in white script on the front, "love bears all things" in pink on the inside, and small raised lettering on the back: "©Samuel J. Butcher." Tush tag: "Applause ®// Div of Wallace Berrie & Co, Inc.// ©1985 Samuel J. Butcher//Product of Korea." $10-$16.

Q

QUAKER OATS CO. - 15" blue velvet bear dressed in red/white checked neckerchief, chef's hat and printed apron: "Breakfast Bear, Aunt Jemima ®." Back tag: "Baby BARE Bear™, ©1984 North American Bear Co., Chicago, IL, Barbara Isenberg Design." A promotion in 1988 offered the bear for $7.99 plus 4 UPCs from Aunt Jemima® brand frozen waffles, French toast or pancakes. Current price $200 each. Aunt Jemima began as a milling company character in 1903, appeared in magazine ads in 1916, and became part of the Quaker Oats family in 1926. *Promotional information courtesy of Jean Laughery. Bear courtesy of Nancy Vanselow.*

RADIO SHACK ® - 10" seated beige bear with silky shaggy fur, plastic eyes, brown velour nose, brown ribbon bow tie, has an AM-FM radio inside him with dials on his tummy. Zipper in bottom allows access to battery pack. Tush tag: "Custom manufactured in China for Radio Shack, A Division of Tandy Corporation, Fort Worth, Texas 76102" Box calls him "Huggable Brown Bear, AM-FM Radio," and says "This furry bear has a tummy full of tunes—plays AM/FM radio just for you!" Bear purchased at a Radio Shack ® store in Florida in 1998. Original price $15.99.

QUAKER OATS CO. (cereals) - 12" seated, honey brown plush bear, cream snout and back paw pads, pink plush circle cheeks, brown velour nose, brown floss mouth, hooded plastic eyes. Furry tufts on bear's head in front of pastel striped night cap with white furry tassel. Pastel striped nightshirt closes with Velcro® strip in back. Velcro ® strips on each paw allows them to assume prayer position. Tush tag: "fp, Fisher Price ®, Division of the QUAKER OATS CO., East Aurora, NY, © Morgan Inc 1985, 1401, Made in Korea." $10-$20.

Left: RAIN MASTER (gutters) - 11" standing cream colored bear with jointed arms and legs, plastic eyes, velour nose, floss paw claws, red neck ribbon. Body tag: "Polyester Fibers// Made in China." Hooded red plastic raincoat with front pockets trimmed in white. One pocket has a picture of Snoopy's pal Woodstock holding an umbrella. "Rain Master// All weather vinyl gutter system" is written across the raincoat's back. Hood snaps on. Tag inside rain coat: "WOODSTOCK// (c) 1965 United Feature Syndicate Inc.// Determined Productions Inc.// San Francisco// Made in Taiwan." The coat may not be original to this bear as it is snug at the neck and the bear seems of more recent vintage than 1965; plus the tag and picture indicate "Woodstock." Possibly a yellow Woodstock bird was the coat's original wearer. $8-$15

Above: Back view of Rain Master bear's raincoat.

RICH'S DEPARTMENT STORE - 21" standing white plush bear, dressed for the holidays in red knit sweater and knit cap with "Richie" knitted into the brim. Paper tag reads "Richie Bear." Body tag: "Made by Chosun International Inc., Seoul, Korea." Purchased on-site at Rich's. *Courtesy of Oneida Callaway, Photo by Mary Ann Callaway Dennis.*

ROSS LABORATORIES, SIMILAC (infant formula) - Newborn babies received this gift package from Ross Laboratories in 1978. Contents: Similac powder, bottle of sterilized water, 2 nipples, and 3" brown bear. The bear appears on the label of the package and on cans of Similac infant formula. Ross is a division of Abbot Labs USA. *Courtesy of Ginny Kreitler.*

ROSS LABORATORIES, SIMILAC - 3.5" seated light brown bear with tan snout, ear linings, and paw pads. Tiny plastic eyes, floss nose and mouth, clip together hands, white nurse's hat says "Ross" in white on a red rectangle. Leg tag: "Rosco ® Teddy Bear// Similac ® with Iron// infant formula." Reverse: "Made in China." Undated. $6-$10

ROSS LABORATORIES, SIMILAC - 6" seated cinnamon brown plush bear, tan ear linings and paw pads, small plastic eyes, white felt snout, black floss mouth, red neck ribbon. Leg tag: "Rosco Teddy Bear, Similac ®, Infant Formula." Tag reverse: "Made in China." $4-$8.

ROSS LABORATORIES, SIMILAC - 6" cinnamon brown plush bear, white snout, cream ear linings and paw pads, plastic eyes, black velour nose, black floss mouth, yellow neck ribbon. Leg tag: "Rosco ™ Teddy Bear, Similac® with Iron, infant formula." Tag reverse: "Made in China." In 1998 these bears were mailed to new mothers. $4-$8.

ROSS LABORATORIES - A trio of 3" seated brown rubber squeak toy bears with painted white snout, tan ear linings and paw pads, molded painted red neck bow. Bottom mark: "© 1985 Ross Laboratories, Made in Hong Kong." (These are not stuffed bears but they wanted to be near their family.) $3-$7 each.

SANDOZ NUTRITION CORP - "fiddle faddle" (glazed popcorn and nut snack) - 11" light gold plush bear with white muzzle, gold cotton bow tie, red plaid shirt. Blue denim overalls say "fiddle faddle" in yellow on the bib. Custom designed clothing is part of his body. Cloth fiddle (often lost) is fastened to paw with Velcro" strip. Hat not original. Two tags: "fiddle faddle ™, ©1986 Sandoz Nutrition Corp" and "Animal Fair by Guy// Eden Valley, MN 55329." In 1985-86 you could get the bear for $8.95 plus $2 shipping and 2 proof-of-purchase panels from Peanut or Almond fiddle faddle. *Courtesy of Jean Laughery, Photo by Vivian L. Gery.*

SATURDAY EVENING POST - 1998, second version of "Post Delivery Boy" bear with his dog (2 piece set), his delivery bag and miniature copy of "Saturday Evening Post." Note his outfit is different from the 1995 version. Made by Douglas Cuddle Toys for their Company Classics Collection. In 1998 Douglas introduced a new 5" miniature Company Classics bear line. The mini newsboy is accessorised with hat, vest, and over-the-shoulder news bag. Issue price of mini $16.50

SANDOZ NUTRITION/ PLANTERS - 11" "fiddle faddle" bear with boxes of his product. His fiddle is missing (confiscated by the neighbors?). *Courtesy of Peggy Monahan.*

SATURDAY EVENING POST - "Post Delivery Boy" Bear, 15", limited edition of 1,200, individually numbered and boxed. Distressed "fur", fully jointed, suede paws, and hand-embroidered nose. He wears Argyle patterned knit vest with coordinated shirt and pants, and an old style billed cap. His accessories include an over-the-shoulder delivery bag and miniature copy of the "Post." This is the first version from Douglas Cuddle Toys, Company Classics Collection, 1995.

For info on our cover gal & guy, see page 2

To continue to receive our catalogs, see order form on page 23.

See pages 19,20,21,24 for Fran Lewis' Bearwares

SATURDAY EVENING POST - 14" Brother Bear takes his little (12") sister with him to deliver magazines in his "little red wagon." A frog in his pocket adds whimsy to the playful scene. The bears are dressed in typical 1950s school clothes. She has red ribbons in her hair, and wears saddle shoes. Limited edition of 1,800 ensembles worldwide. Made by Douglas Company, 1998. Issue price $150.

FAO SCHWARZ - The enormous Truffles Bear who, along with Raggedy Ann, guards the entrance to FAO Schwarz store in Orlando, Florida. *Photo courtesy of Tiffany Chorney, FAO Schwarz.*

FAO SCHWARZ (toy store) - "Truffles, " an 18" seated honey brown shaggy furred bear with velvet back paw pads, quarter-sized amber plastic eyes, dark brown velour nose, and white ribbon bow printed with a smiling clock face, a toy soldier, and a yellow rocking horse. Tush tag has a white rocking horse against a red square and "FAO SCHWARZ Fifth Avenue." Tag reverse: "Made in China." Here is Truffles story: "Truffles the Bear was found in a candy store going particularly wild over the truffles candy…which explains his padding. (We could bearly get him out the door.) Hence the name, Truffles the Bear. Truffles the Bear was 'born' in 1992." Truffles is exclusive to FAO Schwarz and comes in 18 " ($29), 3 ft ($89) and 4.5 ft ($295.00) sizes.

SCHWINN (Bicycles) - 12" girl bear wears pedal pushers, straddles a red 1960s Schwinn Sting-Ray Apple Krate Bike (which comes in its own collector box). She's about to take her cocker puppy along for a run. Made by the Douglas Company, for the Company Classic series. Bear and bike limited to 1,500 pieces, numbered individually. Issue price $119.50, 1998.

THE SEVEN-UP COMPANY - 3" tall round red plastic "Spot" with black oversized sunglasses, round black mouth, black rubbery flexy arms and white gloved hands with a red "Spot." Wind on left side and white metal feet walk. Raised lettering on back: "© 'Spot' Is A Trademark Identifying Product of the Seven-Up Company, Dallas, Texas, 1988. Nasta Ind. Inc. N.Y., Made in China." $5-$10 in working order. The 7-Up name has been used since 1929.

THE SEVEN-UP COMPANY (clear soft drink) - A trio of 7" all white bears, shirts are part of body; plastic eyes, black plastic nose, red floss mouth. Yellow shirts read "Cherry" (in red) "7 UP ®" (in green) with a red cherry on the sevens. Red shirt says: "7 UP ® Gold" with a round circle in light pink flocking. All 3 leg tags are identical: "ACE Novelty Co Inc" on front, and "Los Angeles-Chicago-Seattle, Made in China" on reverse. Stuffed red cotton "UNcola" Spot, with 6.5" circumference, has black felt eyes and black stitched smile, no tag. Seven-Up's Uncola promotion has been ongoing for nearly 30 years. Bears $5-$15 each. Red "Spot" (without tag) $2-$4.

SHIRLEY'S DOLL HOUSE (doll shop, Wheeling, IL) - 7.5" felt bear, stitched claws, nose and mouth, plastic eyes, green neck ribbon, no tags. Removable T-shirt says "I (heart) Shirley's Doll House." *Courtesy of Nancy Vanselow.*

SHONEY'S ® INC. - 8" molded plastic Shoney's Bear bank with coin slot in back of head. Flesh colored face bottom, red shirt and blue pants are painted on. Incised marking on right foot bottom: "Shoney Bear is a Registered Trademark ®, © Shoney's Inc. 1993." On left foot: "Made in China, For Ages Four and Up, Remove Head to Access Coins." *Courtesy of Dianne Rinehart Houser.* $12-$15.

SHONEY'S ® INC. - A quartet of golden brown plush bears with tan snout, brown plastic nose, red felt lining in open mouth, tan ear linings and paw pads. Dressed in removable jeans with shoulder straps and red sweat shirts (with back Velcro® closure) which say (in white) "SHONEY'S ®." Bear second from right is most recent edition. His face is more "little boy" looking. His jeans seem less well made and his tag is different. It reads: (standing bear logo) "Shoney Bear ®" on front and on the reverse: "Imported for Shoney's Inc. by M.O. Money Associates, Inc., Pensacola, FL 32503, © 1995 Shoney's Inc, All Rights Reserved, Made in China." Tags on the other 3 bears read: "Shoney" (bear head) "Bear" on foot tag front and on tag reverse: "Imported for Shoney's Inc by Western Publishing Company, Inc., Racine, Wisconsin 53404, © 1986 Shoney's, Inc., Made in China." In 1988 these bears could be purchased on-site for $4.98 from the restaurant cashier. $8-$15.

SHONEY'S INC. (restaurant) - 11" brown plush bear with beige muzzle and paws, open mouth with red lining, large plastic eyes. Denim jeans with straps and red/white checked shirt are removable. Later bears wore red T-shirts with "Shoney's" name. Tag: "Imported for Shoney's Inc. by L.C.I., Santa Ana, CA, © 1986 Shoney's Inc." *Courtesy of Jean Laughery, photo by Vivian L. Gery.* $15-$20.

SMITHSONIAN INSTITUTION - 14" replica of 1903 early Ideal Toy Company Teddy bear. Rich brown with cream paws, head faces left, frown expression. This Smithsonian specimen is a fine example of the species "Teddy" and is based on the original bear in the collection of the National Museum of American History in Washington, D.C. Tag: "1987 Smithsonian Institution, Determined Productions, San Francisco, CA." Purchased for $20 by catalog order in 1987. *Courtesy of Jean Laughery, photo by Vivian L. Gery.*

SHOPRITE (grocery chain) - 19" standing "Scrunchy", circa 1970s, brown bear with beige snout, ear linings, and back foot pads, amber plastic eyes, round black nose, red felt tongue. His bright yellow turtleneck, long-sleeved shirt is part of his body. The shirt front has a circle with "SHOPRITE" and a black grocery cart against a red background. Untagged. 8" seated "Scrunchy" has same description except no turtleneck collar, and a green ribbon around his neck. Tag and red felt tongue are missing. 19" bear $10-$15 (due to faded shirt and missing tag), 8" bear, missing tongue and tag $10-$18. The 19" Scrunchy sat on grocery store counters in an open box with his shirt logo and "Scrunchy the Shoprite Bear" printed on the box sides.

SHOWBIZ PIZZA PLACE - 3 Showbiz Pizza characters from 1986. In the center 8.5" "Billy Bob" bear holds his wooden guitar. On the right is white furred 10" "Beach Bear." On the left is a non-bear, a 9.5"gorilla named "Fatz." All have removable clothes, plastic hands and feet. Tags: "ShowBiz Pizza Place." Purchased on-site. *Courtesy of Ginny Kreitler.*

SNOW CROP ORANGE JUICE - Circa 1950s, 10" white plush bear, with red circle on his right chest that says "Teddy Snow Crop." A matching Teddy Snow Crop wooden wall clock is 5.5"x8.5". It has chain and metal pendulum. *Photo courtesy of Dee Hockenberry.* Bear $50-$65. Clock $75-$85.

SNYDER'S OF HANOVER (snacks) - 14" tan plush stuffed bear attired in a baker's cap and apron with a twisted pretzel logo, yellow shirt, white pants, and red four-in-hand tie. Bear is called "Bearon VonSnyder, Chief Pretzel Twister." Tag: "TRUDY, Norwalk CT." It was a direct purchase in 1991 for $21.95. Each purchase granted membership to Snyder's Creative Pretzel Eater's Club. *Courtesy of Jean Laughery, photo by Vivian L. Gery.*

STATE FARM INSURANCE ® - A quartet of 11" light brown plush bears with beige snout and back paw pads, plastic eyes, black floss mouth, wearing white T-shirt with red sleeves and neckline and State Farm Insurance logo. Bear far left wears "Good Neigh Bear" shirt and has brown velour nose, body tag date 1993. Bear second from left has a black plastic nose, wears a "Like A Good Neigh-Bear, State Farm Is There" shirt; has a gold oval plastic "SMILE" tag in right ear, body tag date 1990. Bear second from right has brown velour nose and wears identical T-shirt with body tag date 1991. Body tags for these 3 bears are identical except for year: "From the World of SMILE® International. Inc., Commack, NY 11725, ©(year), Made in China" (bear far left), or "Made in Korea" (2 center bears.) Bear far right's leg tag is earlier: "TRUDY, Norwalk, CT USA, Made in Korea, ©1983 Trudy Toys Inc." $8-$15

STATE FARM INSURANCE - 19" plush "Good Neigh Bear." The State Farm Insurance logo appears in red on his shirt along with "Good Neigh Bear." Tush tag: "Animal Fair, A division of Princess Soft Toys, Inc." Tag reverse: "Surface washable// Animal Fair// Mpls, MN 55439// Made in China." This bear was a door prize at a promotional dinner in 1997. *Courtesy of Monica Wilson.*

SUNBEAM CO. - 14" tan plush bear, dressed in typical chef's outfit, white hat and apron that says "Sunbeam Bear, Baker Bear™," red neckerchief. Tag: "Checkmate Promotions Inc., Hackensack, NJ." Offered in 1987, terms of offer unknown. *Courtesy of Jean Laughery, photo by Vivian L. Gery.*

T

TASTEFUL ADDITIONS (candies) - 5" seated black and white plush panda, black felt eye surrounds, black plastic nose, black floss mouth. Round paper ear tag's red heart says: "My Love Is De-PANDA-able." Below the heart in red is "Love's a Jungle." Reverse paper tag: "1 oz Cinnamon Red Hots, Tasteful Additions, Salinas, CA 93901" Original grocery store 1998 price $2.99. Ribbon seat tag: "Steven Smith ©, Stuffed Animals Inc, Brooklyn, NY, Made in China." (Steven Smith makes uniformed bears for many baseball and other teams.)

TED E. BEAR (book and video educational materials) - 6" seated reddish brown bear, ("Professor"?). tan pointed snout, plastic eyes and small black plastic nose, floss smile mouth, dressed in black felt mortar board hat with felt tassel, black rimmed felt spectacle frames, white felt tail coat, black and white striped tie. Leg tag: "Ted E. Bear ™ // and friends (paw print)" Reverse: "Made in Taiwan." 6" seated honey brown bear with tan snout, plastic eyes and small dark brown plastic nose, black felt glasses frames, wearing yellow felt vest and orange bow tie with white dots. Leg tag same as "Professor." Each $5-$12

TED E. BEAR - Trio of 6" golden bears, tan snouts, plastic eyes and tiny black plastic nose, black floss smile, blue felt Greek style hat with black bill and black felt pompom, light green bow tie with white dots, vests of felt, one blue, one yellow. Bear with red bandana (not original) has a green bow tie under his kerchief. Leg tag: "Ted E. Bear ™ // and friends (paw print)" Reverse: "Made in Taiwan" Ted. E. Bear storybook and record (by Starland Music), "The Summer It Snowed in Bearbank" © 1983, is shown with them Bears $4-$10 each. Book $1-$5.

Teleflora

TELEFLORA ® - Cupid, 9" seated cream plush bear, head turned to the right, black plastic heart shaped eyes, black plush nose, black floss mouth, pink felt paw pads, metallic silver quilted fabric wings, silver cord over shoulder holds silver quiver with red plastic arrows. Body and head are wider than "Dream Bear." Ribbon tush tag: "© 1985 Teleflora, Product of Taiwan" $12-$20

TELEFLORA ® - "St. Nick'las Bear," 10" cream bear, head turned to
the right, round plastic eyes, black velour nose, black floss mouth, pink
flannel paw pads, red flannel Santa hat with white trim and tassel, red/
white striped neck scarf, red flannel gift bag slung over shoulder, bear
hugs red plastic chimney flower container. Leg tag: "St. Nick'las Bear
™, ©Teleflora, Product of Korea." $15-$25

TELEFLORA ® - "Dream Bear," 9" seated white bear with head
turned to the right. Plush has lambs-wool look, pink felt paw pads,
black plastic heart shaped eyes, black velour nose, black floss mouth,
cotton night cap printed with red and pink hearts, matching pajamas
with pink heart shaped plastic buttons holding drop seat shut. Clothes
tacked onto him. Arms spring open to hold red plastic heart shaped
flower container (not pictured). Leg tag: "Dream Bear ™, ©Teleflora,
Product of Taiwan." A Valentine's Day item sold from 1988 until 1990
when the supply was exhausted. $12-$20.

Above: TELEFLORA ® - "Valentine Bear," 9" smooth white plush bear,
head turned to the right, 2" stuffed red felt top hat, silver bow tie; red
flannel tail coat is part of bear's body, two white buttons with red "V"
on back of tail coat, black plastic heart shaped eyes, black plush nose,
black floss mouth, arms fit into notches in red plastic heart shaped
container. Ribbon leg tag: "Valentine Bear ™, © Teleflora Inc, Product
of Korea." Body and head are less chubby than Cupid Bear. $12-$20

Left: TELEFLORA ® (floral wire delivery) - 7" seated light brown
plush, cream snout, ear linings, back paw pads, and rounded belly.
Black nose, sunglasses with red hearts on the lenses, white T-shirt
with "Be My Honey" machine embroidered on the front. Tush tag: "A
Teleflora® Gift, Teleflora ®, Los Angeles, Polyester fiber, Product of
China." Offered for Valentine's Day 1995. Behind the bear is the
original magazine advertisement for him. $15, with ad $18.

TENDER SENDER INC. (package wrapping and mailing service) - "Packy" Bear was used in the company's advertisements. The 14" bear, purchased at a Tender Sender store in 1985, came in larger and smaller sizes. Left hip tag: "Mfg. For Tender Sender Inc, Exclusively." His light blue billed cap and tie say "TS." Paper tag: "Animal Fair." *Courtesy of Ginny Kreitler.*

Coming in November 1998

Monkey Island Bears & Friends Presents:

"*Ace*"
Texaco's Flying Pioneer

Please contact your nearest Texaco® dealer to ensure availability in your area.

TEXACO ® - "Ace," Texaco's Flying Pioneer, an aviator with a Texaco star patch on his jacket. His helmet has holes for his ears. Aviator glasses, red neck scarf, and lambs-wool collar, give this beige bear with plastic eyes and brown floss nose a fetching air. Release date November 1998 by Monkey Island Bears & Friends. No further details known at press time.

"TEX"
The Full-Service Bear

TEXACO ® (gasoline) - 16.5" "Tex" The Full Service Bear" recalls "the good old days" of the fifties when the gas station attendant pumped your gas, washed your windshield, and checked your oil, without your asking and (except for the gas, 27 cents a gallon) for free. This white, fully jointed bear wears a red Texaco "Star" on his hat and jacket. He comes in his own collector's box. Box marked: "Monkey Island Bear and Friends, Oklahoma City, OK, 1-800-888-1912." Paper tag: "Tex" the full service bear reminiscent of the 1950s, full-service station attendant." He is the first of a series. *Bear photo courtesy of Tony Tyndall, information courtesy of James Thomas.* First Edition 1997 issue price $39.95, Current price $150-$170. Second Edition 1998 issue price $31.

THOMPSON-HOLLAND - 6" high, 5-1/2" arm span, beige plush bear with red satin neck ribbon, plastic eyes, heart-shaped black plastic nose. Leg tag: "Thompson-//Holland, Inc.//Made in China." Reverse tag: "© 1996 Thompson-Holland, Inc.//Stockbridge, GA." $2-$4.

TIME-LIFE MUSIC - In 1983 Time-Life used two bears to plug their music albums. The bear on the left, with "In the Mood" on his shirt, helped sell Glen Miller albums. The tan bear in the middle, inscribed "Love Me Tender," touted Elvis Presley albums. Both are 12" and made by Trudy. For more information on the right-hand bear see the heading "State Farm Insurance." *Courtesy of Ginny Kreitler.* $12-$25 each.

TOOTSIE ROLL ® INDUSTRIES, INC. (candy company based in Chicago) - "Happy Camper" is a 6.5" bear with his name on his chest, a red fabric cap and backpack holding his outdoor "essentials," a package of 24 Tootsie Rolls ®. Bear colors varied. It was advertised in a mail-order gift catalog in 1993, for $8.98 plus $4.50 shipping and handling. Look also for the girl Tootsie Roll Bear with her head full of Tootsie Rolls used as curlers.

TOYS ™ R ® US (toy stores) - 15" standing bear, grayish brown shaggy pile, white ear linings, small plastic eyes, brown floss nose. Only the forest green velvet over-blouse with attached red and green scarf-type collar is removable. Green long sleeved shirt and red "long johns" are part of body. Red and green plaid skirt matches scarf collar and is attached at waist. Forest green knitted leg warmers are stitched above feet. Back seam tag: "Christopher and Holly, Made Exclusively For Toys R Us, © 1994 Commonwealth Toy, NYC, Made in Sri-Lanka." Evidently Holly had a mate, produced for Christmas sales. $6-$10 each.

TRAPPERS™ - (a line of children's clothes?) 11" standing tan lambs-wool-like fur, tan velour sculpted face and paw pads, 3 medium brown freckles on each side of quarter-sized dark brown felt nose, painted plastic eyes, wide floss smile and paw lines. Wears tan corduroy billed hat and one piece outfit with white cotton knit shirt sewn to tan corduroy pants. "Trappers™" label sewn on shirt front; shirt back ties with white satin ribbon. Leg tag: "100% Acrylic, Made in Korea." $5-$8.

TRAVELODGE INTERNATIONAL - Their signs used to picture a bear in nightcap and nightshirt holding a candle, and when we traveled, my youngest son always wanted to stop at "Teddy Bear Lodge," as he called it. Most of the bear image signs have been replaced with more generic fare. In 1994 Travelodge celebrated their 40th anniversary with a special guest offer: "stay Sunday through Thursday, and receive a gift of 2 Peanuts videos and a stuffed "Sleepy Bear." Current (1998) billboards and TV commercials show Sleepy in a blue nightshirt and cap but no stuffed bears are available. Sleepy shown here in white nightshirt and cap is from 1967, the first year of manufacture. He has felt features, black cotton back paw pads, and no tags, except for "Sleepy" on his hat. $45-$60. *Author's collection.*

TRAVELODGE INTERNATIONAL (California based motel chain) - "Sleepy Bear" has been a symbol for Travelodge since 1954. The bear on California's flag was the inspiration for their little spokesbear. The 12" Sleepy in white nightshirt and cap was the first ever offered, in 1967. The 17" Sleepy on his right in orange nightshirt and nightcap with a black nose is from 1976. The 14" bear on the left is also vintage 1976 (note the red nose and orange feet). Lying down resting is 12" Little Sleepy Brother, another 1976 offering. All have droopy eyelids and "Sleepy" printed on their nightcaps. *Courtesy Ginny Kreitler.*

TWA (Transworld Airways) - Coffee, tea, or bear? This 8" white "TWA Ambassador" bear and his drawstring bag were first sighted in a Flight Catalog and purchased aboard the plane in 1987 *Courtesy of Ginny Kreitler.*

TYCO (building blocks) - 20" "Tyco Teddy" is identified by a patch on his chest. A zippered compartment in his back allows children to store their blocks in him. Purchased at Sears in 1986. *Courtesy of Ginny Kreitler.*

U

UFDC United Federation of Doll Clubs

UFDC (United Federation of Doll Clubs) Bears - Far left:1986, Chicago, Illinois Convention, first Teddy Bear Luncheon, 11"gold colored shaggy pure mohair bear, named "Palmer" for the Palmer House. Jointed, stitched nose and mouth, brown suede paw pads, pupil eyes. Foot tag: "Merrythought, Ironbridge, Shropshire, Made in England." Exclusive UFDC limited edition for luncheon participants. Second from left: 1987, Boston, Massachusetts Convention, 8" golden bear, (53% wool, 47% cotton), stitched nose and mouth, plastic pupil eyes. Tag: "Hermann Teddy Original, Made in West Germany." Made especially for United Federation of Doll Clubs Inc." Exclusive UFDC limited edition of 270. Second from right:1988, Anaheim, California Convention, 9" golden, jointed, "Bully Bear," velvet paw pads dark gold, stitched paws, nose, and mouth, plastic pupil eyes. Dressed in sweater with Peter Bull pin. Tag: "House of Nisbet, Ltd., Made in England." Exclusive UFDC limited edition of 310. Far right: 1989, UFDC 40th Birthday, St. Louis, Missouri, 9" black genuine mohair bear, jointed, black pupil-less eyes, floss stitched nose, felt paw pads, red neck bow . Tag: "Steiff, Made in Germany." Exclusive UFDC limited edition of 360. *Courtesy of Nancy Vanselow.*

UFDC 1988, Los Angeles, California Convention. The 1987 Hermann Teddy bear was auctioned off with a Mickey Mouse wardrobe made by Shirley Buchholz and Joyce Kintner. In the basket: leather cowboy suit, chaps and vest with pistol; engineer's outfit with lantern and a tiny train; clown hat and collar, nightgown and cap with hurricane lamp, mandolin, miniscule Holy Bible. Wicker table, plaid tablecloth and tea set. Mickey Mouse ears and leather suitcase. *Courtesy of Nancy Vanselow.*

Rear view of 1987 Hermann Teddy dressed as Mickey Mouse showing his curled tail.

UFDC Convention Luncheon Bears -Far left:1990, Washington, DC Convention, 9" gray brown bear, jointed, hump back, stitched floss black nose and mouth, black eyes, beige stitched paw pads, not dressed. Tag: "Bearly There Co., Westminster, California, by Linda Spiegel-Lohre." Exclusive UFDC limited edition of 410. Second from left, 1991, New Orleans, Louisiana Convention, 10" light gold mohair bear, "Francois," jointed, hump back, black eyes, black stitched nose and mouth, beige felt paw pads, black beret, white Peter Pan collar, red neck bow. Signed tag: "By Kathi Clarke." Exclusive UFDC limited edition of 400. Second from right, 1992, San Francisco, California Convention, 8" beige mohair bear, "Sue-Ling," jointed, suede paw pads, black eyes, stitched nose and mouth. Dressed in Oriental outfit of black pants and red/gold print top, coral cord closure, black "beanie" hat. Designed by internationally known designer Jenny Krantz, made by Owassa Bear Inc., Owassa, Michigan. Exclusive UFDC limited edition of 310. Far right, 1993, Chicago, Illinois Convention, 9" golden mohair bear, "Isabel," jointed, pupil eyes, brown stitched nose and mouth, maroon ribbon headband says "Isabel by John Axe." Teddy bear passport exclusive for Isabel. It is attached to her wrist. Tag: "Merrythought Ltd, Ironbridge, Shropshire, England." Exclusive UFDC limited edition of 300. *Courtesy of Nancy Vanselow.*

UFDC Convention Luncheon Bears: Far left, 1994, Atlanta, Georgia Convention, 7" light gold mohair bear, "Regina," jointed, brown stitched nose and mouth, tan pupil eyes. By John Axe, Made by Alpha Fargill in England. Exclusive UFDC limited edition of 313. Second from left,1995, Philadelphia, Pennsylvania Convention, "Remember the Ladies" theme, 9" light brown genuine mohair bear, black button eyes, stitched black nose and mouth, beige felt paw pads. Mauve pink ribbon chest banner: "Remember the Ladies - Steiff - UFDC 1995." Tag: "Antik Teddy bar, 26, genuine mohair." Tag: "Made by Steiff Germany Ltd." Exclusive UFDC edition limited to 365. Second from right, 1996, Dallas, Texas Convention, "Beautiful Children" theme, 13" golden mohair bear with growler and hump back, black button eyes, felt paw pads, black stitched toes, nose and mouth, Tag: "Steiff, Made in Germany." Far right:1997, Anaheim, California Convention, 9" white mohair, "To Have And To Hold Bridal Bear" with squeaker, white suede paw pads, black eyes, nose, mouth, and stitched toes. Wears head wreath of pink flowers and leaves with pink veil and bouquet. Exclusive UFDC limited edition of 365. *Courtesy of Nancy Vanselow.*

NO PHOTO

UFDC Convention Luncheon Bear - 1998 Convention, New Orleans - Small dark bear named "Napoleon," made by Kathi Clarke, limited edition of 50. *Information courtesy of Nancy Vanselow.*

U.S. MARINE CORPS - 11" seated white plush bear with plastic eyes and nose, and flat tail. He wears a red T-shirt with "United States Marine Corps" and the USMC emblem of eagle, world, and anchor in gold. Body tag shows a bear logo, "ASI 62960, Made in Indonesia." Purchased from the Donlen Company catalog (exclusively items of interest to Marines) for $18.75 including shipping. This little guy has a companion whose red T-shirt reads: "I (heart) My Marine."

U.S. AIR FORCE - 20" dark brown bear, khaki nylon flight suit, camouflage billed hat, black plastic high top boots. "Bear Forces of America" patch. US flag on sleeves, "US Army" and wings patch. Zippered pant legs, pockets, front, and sleeve. Metal dog tags around neck. Neck tag: "Bear Force of America ®, © 1989, Ira Green Inc., Made in Korea." *Courtesy of Nancyann Eckhart.* $35-$40.

U.S. POSTAL SERVICE - 20" brown bear, lighter snout. Dressed in skirt, blouse, necktie, black plastic boots, zippered sweater with two pockets and "Letter Carrier" patch. Billed hat with ear holes has patch with Eagle and "U.S. Mail." Mailbag has leatherette strap and "U.S. Mail" patch. Neck tag: "Patriot Bear ™" ©1986 J.J. Wind, Inc. under exclusive license of the United States Postal Service, Made in Korea." Paper tag: "Patriot Bear ™." Inside: "Patriot Bear ™, The service oriented 'Patriot Bears ™' are dedicated with pride and affection to the service people in the USA." Tag back has washing instructions. There also was a male mail carrier bear. *Courtesy of Nancyann Eckhart.* $45-$60.

USA, UNCLE SAM - A bevy of proud Uncle Sam bears lined up to watch the 4th of July parade. These are fun to find, usually inexpensive, and work well as holiday or table decorations. The tallest bear is 14". Note the white goatee on Uncle Sam bear on the left middle ladder step. He is 9" by Dan Brechner. Bears $2-$10 each.

Left: USA, UNCLE SAM - Uncle Sam and our flag are instantly recognizable symbols of our country. They advertise our democratic government, and the "land of the free and the brave." This little 3.5" pink clip-on bear sports both an Uncle Sam hat and Old Glory on the back of his red velour vest. Note the hearts on his paw pads. He seems to be saying "I Love America." (Me too!!) $5-$8.

Above: Back view of pink Uncle Sam bear.

VICTORIA'S SECRET - 12" seated pink plush bear, plastic eyes, brown velvet nose, black leatherette paw pads. Tag: "© Gund Inc 1992, Made in Thailand." Although this bear is missing its plastic heart and paper "Victoria's Secret" tags, it's a twin and probably was a Victoria's Secret bear. They were offered in a variety of colors. However, without the tags the value is lowered to $8-$15. Gund celebrated its 100th anniversary in 1998.

VARSITY ™ (children's clothing line) - 30" (standing) bear wears removable red sweater proclaiming him a "Varsity ™" teddy. Sweater has black cuff stripes. Tag: "© 1985." Purchased at J.C. Penney store in 1987. *Courtesy of Ginny Kreitler.*

LILLIAN VERNON CORP. (mail order catalog) - 14" seated black and white panda, white tail, plastic eyes, black plush eye surrounds, black plastic nose, red felt tongue, humped back, front paws rest on back paws. White ribbon leg tag: "Lillian Vernon Corp., Mount Vernon, NY 10550, 100% wool, Made in China." $8-$12.

VICTORIA'S SECRET (lingerie) - 12" seated white plush bear, black leatherette paw pads, brown velvet nose, plastic eyes, blue and green plaid bow, heart-shaped plastic disk tag: "Victoria's Secret" Tag reverse: "© Gund Inc.1992." Paper tag: "Victoria's Secret." *Courtesy of Helen B. Evans, Billie's Emporium.* $10-$20 with all tags.

Wal-Mart

WAL-MART STORES, INC. (discount store) - 14" (standing) brown plush bear, whose back paw pads are blue satin machine embroidered in white with "Liberty Bear" on the left foot and "1992" on the right foot. Ears are red with white dots, plastic eyes, velour nose, and red/white/blue neck ribbon. Tag: "Marketed by Wal-Mart Stores, Made in USA". Flag is not original with bear. *Courtesy of Thada Swanson Collection.*

WAL-MART STORES - 11" white plush bear, red ear linings, navy blue back paw pads, plastic eyes, black velour nose, black floss mouth. Her navy blue with white stars print top, red sleeves, and red/white striped pants are part of bear's body. Red and white striped gathered headband with bow of star print, navy ruffles on sleeve edges, red satin neck ribbon, red ribbon on left foot says "1995." Red/white/ navy paper tag reads: "Liberty Bear, Printed in the U.S.A." Paper tag reverse: "© Dan-Dee International Limited, 106 Harbor Drive, Jersey City, NJ 07305, Made in China, Partially stuffed and hand finished in U.S.A." Tush tag: "Dan Dee, Made in China." (Dan Dee has been making cloth dolls at least since the 1970s.) Although no tags say "Wal-Mart" the 1992 "Liberty Bear" was from Wal-Mart. This Liberty Bear girl had a boy Liberty Bear companion. $15-$20 each.

WAL-MART STORES, INC. - 7" seated golden brown bear, red velour nose, plastic eyes; ear linings, back paw pads and cone are red satin with white hearts, "ice cream" is pastel satin. Cone tacked to his hands. May have been a Valentine's Day offering. Tush tag: "Marketed by Wal-Mart Stores, Inc., Made in China." $5-$8.

WAL-MART STORES, INC. - 11" bright pink (almost maroon) plush bear with plastic eyes and nose. Who could resist this happy color? Leg tag: "Marketed by Wal-Mart Stores, Inc., 702 SW 8th St, Bentonville, AR 72716, Made in Korea." Easter egg is not original to bear. $6-$12

WAL-MART STORES, INC. - 15" honey beige plush bear with beige snout and paw pads, plastic eyes, velvet nose, embroidered mouth, plaid neck bow, no tail. Leg tag: "Made in Taiwan, contents polyester fiber, © Wal-Mart Stores, Inc." *Courtesy of Peggy Monahan.*

WAL-MART - 9"seated white Valentine bear, red heart nose, paw pads red with white dots, plastic eyes, holds (sewn to hand) red heart with white hearts and lace trim on it. Paper tag: "Be My Valentine." Tush tag: "Marketed by Wal-Mart Stores Inc., Made in China." *Courtesy 2nd Chance Home Furnishings.*

WEHRENBERG MOVIE THEATERS - 8.5" chocolate brown bear, yellow snout and ear linings, plastic eyes, brown plastic nose, orange neck ribbon printed with "I.Wehrenberg." Side tag: "Animal Fair." Purchased at the movie theater in 1986. *Courtesy of Ginny Kreitler.*

WAL-MART - 6" seated light pink bear with black plastic eyes and nose. Tag: "Made in Taiwan, Wal-Mart Stores Inc., Bentonville, AR." The rocking horse is not original to the bear. $4-$7.

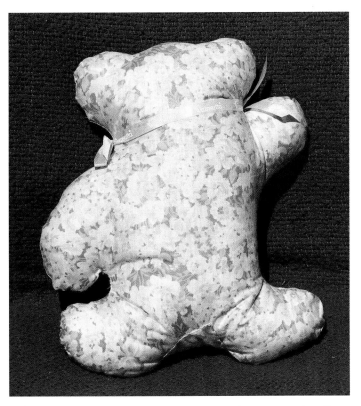

WAMSUTTA ® (fabric, sheets) - 12" pastel floral print cut-and-stuff bear, light blue ribbon around neck says "Wamsutta ®" in white. Possibly was an in-store display item. $4-$8.

WEIGHT WATCHERS ® - 11" seated all white plush bear with head turned to the right, nickel sized plastic eyes, black plastic nose, cream snout indented for mouth, cream back paw pads with black floss toe markings, hot pink arm band, hot pink headband with "Weight Watchers" in white.(This must be her "before" picture.) Hot pink paper tag: "Weight Watchers ®" on front. Inside of folder tag reads: "Hi! I'm W.W. Cuddles! I'm a gift to you from Weight Watchers. I can give love and encouragement to you in exchange for a hug or pass me on to your favorite little person, so they can think of you whenever they give me a hug. Remember, I'm extra soft so I need lots of extra love!" Back of paper tag: "Printed in Hong Kong." Ribbon tush tag has a green tree and "Outdoor Recreational Products, San Diego, California 92111" Tag reverse: "Manufactured by Determined Productions Inc, San Francisco, CA 94126, Product of China."   W.W. Cuddles was used as an incentive for losing weight, attending meetings, door prizes, recruiting new members, could be sold or raffled, dispersed in any way the franchise decided upon. Available from 1988-1993. Current value $18 to $25.

WEIGHT WATCHERS ® (weight loss program) - 10" seated brown plush bear with cream snout, ear linings, and paw pads, brown plastic nose, amber plastic eyes, brown floss mouth. Yellow long-sleeved shirt is part of body; front of shirt has red heart over "Weight Watchers ®" in black script. Probably made for the Canadian market as leg tag is in French and English: "Made by ONT Reg No., Taiwan." $12-$18

WENDY'S (fast food restaurant) - Set of four 7" light tan plush Furskin bears in removable clothes, tan plastic boots, and felt hats with ear holes. Often found without hats. From left to right: Boone, Farrell, Dudley, and Hattie Furskin ™. She wears peach cotton panties under her dress.. Story says she got her blue ribbon at the county fair. Tag: "Graphics International Inc, Kansas City, MO,© 1986 Original Appalachian Artworks, Inc." Paper tag: "Wendy's Old Fashioned Hamburgers, © 1986" Reverse says "Happy Holidays" with a border of red poinsettias and green garland. This was an in-store promotion for the holidays, 1986, price $2.99 each. You could buy one each week for the four weeks before Christmas. *Promotional information courtesy of Jean Laughery. Photo by Mary Ann Callaway Dennis, Courtesy of Oneida Callaway.*

WHATABURGER ® (burger restaurant) - Five 8" stuffed bears, Yellow hatted farmer in denim overalls, white bear with "State College U" shirt, and white felt hat, and beige bear with orange hat and shirt all have the same tags: "Whatabear...g-r-r-r™" on tag front and on reverse "© California Stuffed Toys, Los Angeles, Handcrafted in Korea." Beige bear in yellow "Go Team Go" shirt and brown bear with sewn on jeans and red bandana have matching tags: "Whatabear...g-r-r-r-" on front and back "© CALTOY, Inc, Los Angeles, Handcrafted in Korea." (These were promotional bears and should not be confused with designer "WHATABEAR" bears of Seal Beach, California.) Bears found in Pensacola, Florida, where a Whataburger restaurant is located. The Whataburger/Dr. Pepper cup shown was a thank you to Desert Storm veterans "for a job well done." Bears $4-$10 each.

HELEN WOLF (women's clothing store) - 4.5" bear, neck ribbon says in red "Christopher," has silver hang loop on head. Christopher is a line of women's nightwear. Bear was an in-store promotion for Christmas 1986. *Courtesy of Ginny Kreitler.*

ZAYRE CORP. (department store) - 16" standing tan plush bear, light yellow snout, ear linings, and paw pads; plastic eyes, black plastic nose, black floss mouth, red felt heart centered on chest, red satin neck bow. Leg tag: "ZAYRE CORP., Framingham, Mass. 01701, Made in Korea." Zayre stores were popular in Florida, Maine, and other states during the 1960s and 1970s. $12-$18.

Trademark Bears

Trademark bears are those which are unusual and have been trademarked because of a special feature, for example roller skates, writing arm, tooth fairy pouch. Often no other tags or markings, except for the trademark, appear on the bears, making it difficult to date them.

PINEAPPLE BEAR ™ - 12" gold printed cotton "pineapple" with green felt leaves. Unzip the pineapple and the 10" seated gold plush bear comes out. Paper tag reads: "Pineapple Bear ™, G-B Pro-lines, Honolulu, Hawaii 96819, Made in China." *Courtesy of Nancyann Eckhart.*

 10" plush Pineapple Bear™ emerging from his cotton pineapple encasement.

BATBEAR - 12" standing bear with gray plush legs and front paws, brown felt back paw pads, blue plush head, white snout, pink plastic nose, plastic eyes, purple ear linings, and a purple flannel-like shirt (part of his body) which says "BATBEAR" in yellow against a white bat shape. Batbear wears a blue velveteen cape. No tags or marks of any kind. Was he unauthorized? With him are a 6.5"plastic "Batman Spin Pop Candy" container with a black fabric cape, "™ and © D.C. Comics, U.S. Patent No. 5,209,692" And a 3" purple Batman car with movable black tail, bottom marked "©1991 D.C. Comics Inc, © 1991 McDonald's Corp., China." Batman, Bob Kane's comic book super-hero, first visited Gotham City in 1939. Batbear $10-$15, candy container $5, Batcar $3-$10.

10" golden Pineapple bear with plastic eyes and nose. His bow tie, ear linings, and front paw pads are a print of "Aloha" and rainbows on a white background. *Courtesy of Nancyann Eckhart*. $15 up.

Far Right: SKIPPY SCRIBBLES - 16" seated tan bear with pale rust snout and foot pads, dressed in a removable red shirt and red/white striped overalls which drop in back for battery pack access. Marked on overalls "Wonderama ®", buttons marked "Skippy Scribbles." This mechanical bear has a long (left) bendable arm of wood, hand can hold crayon or pencil. Control mechanism is a wheel in the back of the neck. No tags. *Courtesy 2nd Chance Home Furnishings.* $15 up.

ROLLIE™ - A 12" brown bear on 2" roller skates. The base of the bear is a vinyl doll body, with plush fur covering only the head, chest, arms, back, and a strip on each ankle. Beige plush snout and ear linings, black plastic nose and eyes with white felt surrounds, red felt tongue, brown vinyl "people" hands. His overalls have two felt buttons in front. A 5" Velcro® closure in the back allows access to battery pack. On the bib pocket is machine embroidered "Rollie ™" No further tags or markings. $5-$12

Right: "SNUGGLE BUGGLES" - "Snuggle Buggles" is a brown bear relaxor massager "filled with herbal essences for aroma-therapy, creates a calming, stress-reducing effect, 250 JBR" 1998 issue price $19.99.

TEDDY GRAM - 14" standing white plush bear with tail, plastic eyes, red velour nose. Bow tie, ear linings, and paw pads are a red cotton print with white hearts. Plastic tag in left ear "Compliments of Applause." Tush tag: "Applause, Division of Wallace Berrie & Co. Inc, Woodland Hills, CA, Made in Korea, © 1987 Applause." Paper folder tag: on front "Teddy Gram, This Bears My Love" and a bear dressed like an old fashioned telegram delivery person. Inside the folder, printed in red: "Teddy Gram, To (space), From (space), Received message * Stop * Terrific Teddies Say It All * Stop * Send Your Love With Teddy Grams * Don't Stop* Signed Teddy." $15-$18

WINDSHIELD SCRAPER MITTEN BEAR - 9" white plush bear with arms and tail, plastic eyes and nose, red scarf with green fringe, red and green knit cap with ear holes and hole on top where black plastic windshield scraper sticks through. Lined in heavy cotton knit. Although this little bear appears to be commercially made, neither he nor the scraper have any markings except for a lone ® mark on the scraper handle. *Courtesy of Muriel Hoffman.* $4-$8.

TOOTH FAIRY BEAR ™ - 8" seated pale peach plush bear with black eyes, and white snout with beige velour nose, floss mouth. She holds a 3"x2.5" white pouch with "Tooth Fairy Bear ™" machine embroidered on it in bright pink. Tush tag: "Animal Fair ®™ Product, Minneapolis, MN, Product of Korea." Tag reverse has contents, washing instructions, and "© Animal Fair, Inc." $10-$15.

2. Other Ad Bears to Hunt For

The following list contains advertising bears the author has learned about but was unable to photograph. It is provided to add to your knowledge and wish list. You are sure to have or find many other different advertising bears than those listed or pictured here. Good luck in your hunting.

Where the bears are pictured in another publication they have been assigned code letters:

(H) = *Hake's Guide to Advertising Collectibles, 100 Years*
(R&S) = *Advertising Dolls,* Joleen Robison & Kay Sellers
(Z) = *Zany Characters of the Ad World,* Mary Jane Lamphier
(TBf) = *Teddy Bear & friends*
(GMA) = collector Gloria McAdam (no photo)

A&W Root Beer - 9" plastic figural sipper bottle, 1994 (Z)

Ace Hardware - 16" seated, c. 1994, knitted cap says "Ace"

Aim Toothpaste - 12" jointed, plush, in NFL jersey, 1984

American Express - 9" seated plush, "Do You Know Me?" (GMA)

Bass Pro Shops - 30" plush "Hairy" bear in camouflage suit & hat with Bass Pro Shop patches, Kamar tag. (TBf 3/98)

Bear Brand Hosiery - 9" lithographed cloth Papa, Mama, Big Boy Bears, c. 1920s (R&S)

Ben & Jerry's Ice Cream - Plush bear by Douglas Cuddle Toys (1st edition)

Bernstein Chiropractic - 15" plush bear "I Love Chiro-practic" (GMA)

Black Forest Gummi Bear - 19" plush, 1992, name on his tummy (Z)

Blue Diamond Nut Co. - plush bear (GMA)

Bosco Bear - 8" glass and plastic figural bank, c. 1940s (Z)

Bradlee Stores - 14" gold plush "Bruff" made by Animal Fair (R&S)

Carrows Restaurant - 11" bear in blue jeans and red shirt, looks like Shoney's bear (GMA)

Clairol Bear Twins - 15" plush with vinyl masks, 1958, "Honey Blonde" with red ribbon is smiling, brown crying bear wears blue ribbon (R&S)

Clorox - plush bear (GMA)

Close-up Cuddle-up Bears - 10" dark brown plush pair, hugging, 1977. He wears blue felt baseball cap, she wears pink ribbon. (R&S)

CoCo Bear - 12" plush, 1988, shirt says "88", Coco Wheat Cereal, Little Crow Foods (Z)

Crocker Bank - 17", 1975 (in bear price guide) (GMA)

Dole - 10" yellow plush Bananabear ™ (holds banana), Bananimal, 1983, Trudy Toys (Z)

Domino Sugar Bear- 15" plush, 1975, original price $2.98, Bow tie says "Domino Sugar" (R&S)

Foozie Bear - 15" (Sesame St. character), plastic hat and scarf. (GMA)

Gen'l Foods - 12.5" rust plush Post Cereal's Sugar Bear, felt face, aqua body shirt, 1972-76 (R&S)

Hamm's Beer Bear - 8" vinyl black/white display piece with one foot on a pedestal (R&S)

Hamm's Beer Bear - Black/white plush with radio inside, accessible thru back zipper (R&S)

Hamm's Beer Bear - figural ceramic decanter 3.5" x 4.5" x 11", 1973 (H)

Harley-Davidson - 3" hand painted resin figures, numbered limited edition 1998, "Mr Fix-It," "With You I Will Go Anywhere," etc., Miles Kimball 1998 Christmas catalog

Hickory Farms - 12" seated plush bear

Honey Graham Bear - 11" brown plastic figural bank, 1990, Limited Edition by Street Kids (Z)

Huggies - 15" plush bear, "I Love Huggies" on chest (GMA)

ICEE - Pez container

Kellogg's - 1925 Goldilocks (14") & 3 Bears, Papa & Mama (14"), Johnny (12") (R&S)

Kellogg's - 1926 Goldilocks (13") & 3 Bears, Papa & Mama (13"), Johnny (10") (R&S)

Lanz Sleepwear - 12" white plush bear in Lanz blue flannel nightgown (matches child's) (TBf 3/98)

Listerine Antiseptic - Mama, Papa, and Baby Panda, 1971 (R&S, no photo)

Lyon's Restaurant - plush bear (GMA)

Macy's Dept Store - plush bears (GMA)

Maxwell House Coffee - Koala, 1971 (R&S no photo)

Mervyn's Dept Store - 15" Muffin Family Bear, overalls and plaid shirt (GMA)

Mervyn's Dept Store - 9" Merry Muffin (GMA)

Mervyn's Dept Store - 9" bear with diaper and cap (GMA)

Mott's Apple juice/sauce - 9" brown plush grizzly on all fours, c.1993

Oreo Cookies - plush bear by Douglas Cuddle Toys

PG&E - 5" Panda, "Thank you for volunteering" (GMA)

Robinson's Dept Store - plush bear (GMA)

Snow Crest Bear - pint sized glass figural bank, c. 1940s (Snow Crest Beverages, Salem, MA) (Z)

Snow Crop Orange Juice - 8.5" plush hand puppet Teddy Snow Crop with vinyl mask face, c. 1950s (R&S)

Snow Crop Orange Juice - 10" plush Teddy Snow Crop matches hand puppet, c. 1950s (R&S)

Snow Crop Orange Juice - 17" plush Teddy Snow Crop pajama bag, mid-1970s (R&S)

Snow Crop Orange Juice - 14.5" terry cloth Teddy Snow Crop, felt features, 1972. (R&S)

Tastee-Freez - "Bear of a Burger" plush and felt hand puppet with open mouth, mid-1970s (R&S)

Teddy Bear Museum of Naples - a wide selection of T-shirt and banner bears in their gift shop.

Teddy Graham Bear - 11.25" plush, Nabisco, 1990, dressed in jacket, shoes, sunglasses (Z)

Tide - 27" brown/gold plush bear, blue neck ribbon, 1971 (R&S, no photo)

Tide - 24" brown/tan plush, red neck bow, 1976/77 (R&S)

Tonka - 10" plush, "Tonka" on tag (GMA)

Tootsie Roll - plush girl bear with head full of rollers (rollers=Tootsie Rolls)

Travelodge - "Sleepy Bear" Jim Beam liquor decanter

Travelodge - 5 ft "Sleepy Bear" (blue outfit) cardboard display figure, c. 1998

Victor's Eucalyptus Cough Drops - 17" furry plush, Victor Koala Twins, hugging, 1975 (R&S)

Victor's Eucalyptus Cough Drops - 24" furry plush Koala. (R&S)

Von's Grocery - plush bear (GMA)

Wells Fargo - 8" seated plush bear, "Wells Fargo" on shirt (GMA)

Lots of Beary Good Stuff

Many people search for "bear wares" as well as the bears themselves. The following six chapters should give you lots of ideas of what to look for when you're shopping, traveling, or on thrift shop "junking" junkets.

3. Containers

As you'll see by the following photographs, bears adorn a varied range of containers. Mugs, glasses, jelly jars, baby bottles, tins and boxes, all are represented, with bears "pushing the products" of a wide variety of companies.

Reverse side of the A&W glasses.

5.5" A&W's Great Root Bear hugging the glass and looking over his shoulder, and 6.5" The Great Root Bear in a turtleneck sweater with A&W logo. $5-$10 each.

9" sipper bottle with A&W Great Root Bear holding a frosty A&W mug, and 6.5 plastic mug showing Root Bear carrying a foaming over A&W mug. $5-$10 each.

Reverse side of sipper bottle and plastic mug. It seems strange that a root beer stand would be promoting Pepsi. Mug says "A taste to remember, A&W®, Restaurants"

6.5" barrel shaped white A&W sipper with black cap, orange straw, white handle. It says "A&W Hot Dogs & More! A&W." $5-$10. Tiny pompom Root Bear tree ornament sits beside barrel.

9" bright pink sipper bottle with black cap (missing straw) says "Alaska" vertically, on the sipper cup bear is holding. His printed T-shirt says "State Bird, Alaska" and shows a huge mosquito. He wears "cool" sunglasses and shorts. The drawing of the bear is signed "D Sims." "AK" is the only marking on the bottom of the bottle. $3-$5.

4"x3.5" oval blue plastic container says "Bernie the Bear ™, © 1986 CWSF Ltd, All rights reserved" on a paper label on the side. In raised lettering on the bottom "Bernie the Bear ™, © 1984, CWSF Ltd, Made in Hong Kong, 012A." (Notice the years are different.) Contents or identity of Bernie the Bear is unknown. $1-$2.

4" ceramic mug and ad from Bear Mechanical Auto Repair in Bonners Ferry, Idaho, 1994.

Back view of lidded "Celestial Seasonings" ceramic cup shows 2 little bears sleeping.

Front view of lidded ceramic cup marked "Celestial Seasonings Sleepytime Herb Tea" with Sleepy Time Bear snoozing in his chair wearing a red nightcap. *Courtesy of Jean Laughery.* $15-$25.

6.5" Coca-Cola polar bear lidded plastic cup. "© 1993 The Coca-Cola Company." $3-$6.

7.5" yellow plastic mug has an acorn before "Charter, Chubby Chugger (bear), Try A New Twist, Refills anytime 59 cents." Molded into the bottom of mug "46 oz Monster Mug, World's Largest." Purchased in Florida. $1-$5.

Reverse of Coca-Cola polar bear cup "Fill it up at Denny's, 49 cent refills on next visit." Offer expired 12/30/94.

Three 8"x4.25" Coca-Cola collector tin hinged-lid boxes. One shows a lounge-about bear in sunglasses. The second a Coca-Cola basket ball player in red Coca-Cola shirt, and the third a baseball playing bear. Future problems of dating are foreseen due to the only markings being on the paper label on the box bottom. It is written in French, Spanish, and English. "©1997, Coca-Cola, Made in China." $5-$10 each.

Hamm's® Beer stein, 6" ceramic painted in soft colors. Front side shows Hamm's Beer Bear skiing downhill. Stein bottom: "Pabst Brewing Company, Hamm's Beer Stein, Limited Edition, 205." $15-$35.

Reverse of 6" Hamm's Beer stein looks as if Hamm's Beer Bear is accepting a golden trophy. From 1844 until 1889 the original brewery was known as "Best," the name of the founder. In 1889 it was renamed Pabst, and in 1906 "Blue Ribbon" was added to the label. From 1882 until 1906 they had tied an actual blue ribbon around the bottle neck..

4" round pink plastic box with raised brown plastic bear on the lid. Lid top says "Heritage Fun Club." The only other marking is "Made in Hong Kong" in raised letters on the box bottom. $1-$5.

Hershey's (red) Kiss tin, 7.5" tall x 8.5" wide. Chocolate ingredients are printed on the tin's bottom, along with "Mfd. by Hershey Foods Corporation, Hershey, Pennsylvania, 17033, U.S.A., © 1993, HFC." $2-$6.

Right: 9" ICEE sipper stands next to graduated paper cup sizes. All show ICEE Bear drinking an ICEE with a long straw. The sipper and clear straw both have red caps. The bottle inscription: "ICEE, coldest drink in town ™." Raised lettering on sipper bottom identifies "Countryside Products, Pickerington, Ohio 43417." Sipper bottle $4-$8.

3.5" ceramic mug with smiling beige bear against a red background, stripes at top and bottom say alternately "Hickory Farms" and "Hickoryville." A paper tag on the bottom reads "Hickory Farms of Ohio" and "1987." $1-$4. Hickory Farms at one time offered a plush Hickory Farms bear.

NO PHOTO

In 1998, ICEE changed to styrofoam cups with a clear plastic dome cover. The cups depict ICEE bear, in dark glasses, blue jeans, and red sweatshirt with a white initial "I," skateboarding on an ice flow. The cup reads "Too Cool" and around the top and bottom borders "K-Mart Café." The back of the cup advertises Coca-Cola. The cup bottom reads: "Sweetheart."

5" brown ceramic honey pot shaped like a little brown bear (whose head comes off). Incised in bottom "© Al H., Houston Foods, 1982." Even non-bear collectors (there are a few) might easily fall for this little cross-eyed bruin. $8-$18.

5" tall round tin depicts seven bear children lined up to talk to Santa Bear. If the tin loses its bottom paper tag "MARS ™" it will be unidentifiable as an advertising tin. $1-$5.

121

3.5" white ceramic mug with gold top rim and gold vertical stripes. A gray bear hops over the word 'Obie', "The Birthing Suites, East Pointe Hospital." Purchased in Florida, but hospital and location is unknown. $1-$5.

Sea World plastic sipper cup and plastic mug. They show a seal in the sailor outfit that a costumed bear wore at Sea World. $1-$5 each

8" round tin with Santa Bear inside a wreath holding a "Happy Holidays" banner. On the tin bottom in raised lettering "Perkins Family Restaurant Bakery." $2-$8.

4" white plastic "Shoney's Classic American Food" cup. It pictures Shoney Bear dancing with his friends whose names are written around the cup "Shoney Bear, Rita, Penny, Bunny, Grandpa." Grandpa with his white handlebar mustache and pink guitar gets my vote for cutest. On the cup bottom in raised letters "Top Rack Dishwasher Safe" and the recyclable triangle. $2-$8.

7" clear molded glass figural jar of seated bear. Molded into the front "Skippy." Printed on blue plastic lid is "Skippy® Creamy Peanut Butter, (list of ingredients), Best Foods, CPC® International Inc, General Offices Englewood Cliffs, NJ, 48 oz." A paper tag on the bear's back suggests "Best when purchased by Oct 16/90." The 6" clear molded glass figural bear jar with red plastic lid has a molded collar and 4 buttons. His only marking is molded at bottom back "© Kraft General Foods Inc., © 8 R 9771 18." $5-$12 each.

A pair of 4" plastic mugs. On red mug is "Bubi Bear, © 1971, Hanna-Barbera Productions, Inc." On the yellow cup is "Square Bear, © 1971, Hanna-Barbera Productions, Inc." $3-$12 each.

Side view of panda "Tang" shaker, showing the product name.

Front view of 7" plastic Panda shaker for mixing "Tang" (orange flavored powdered drink developed for the Space Program). Instructions for mixing are given on the back, plus "Do not put in dishwasher" and "General Foods Corporation, PO Box 6-QTJ, White Plains, NY 10625 U.S.A." Missing lid. $10-$15.

3.5" heavy white ceramic mug pictures a brown bear wearing antlers standing next to a sign that says "Do Not Feed The Bears" The bear is holding a sign "I am Not a Bear." "Trust Me." It seems a very Yogi Bear kind of thing to do. Marking on mug bottom: "Brinker Ink, Oakland, CA 94611, © 1988, All Rights Reserved." $4-$10.

4. Signage

Bears on signs have ancient origins. As you travel you're bound to come across appealing or interesting bears on billboard, truck, and business signs. Capturing these advertising bruins for your collection is as easy as shooting their photos, and creating an album similar to a "grandmother's brag book." Be sure to take notes of the place and date you found the sign.

Good Time Ice - Relaxing bear painted on the back and sides of a delivery truck at Hershey's Park, Hershey, Pennsylvania, May 1998.

Hamm's Beer - A 5" tall black and white styrofoam store display of the Hamm's Bear holding a tray. (1992 value $100.) Hat has been added. Found outside an antique shop in Brookings, Oregon, 1994.

ICEE - Lighted ICEE sign at food court, Town Center Mall, Port Charlotte, Florida, 1998.

Koala Mattresses - Billboard sign of climbing koala in nightcap discovered in Pensacola, Florida, 1990.

Polar Delight - Polar bear dressed in hula skirt and lei enjoying a "frozen shave ice" snow cone, painted on the back and sides of a sales-booth truck in Pensacola, Florida, 1990.

5. Paper Advertising

To get the word out, it's a "bear necessity" to advertise. Newspaper and magazine ads portraying bears are inexpensive and informative to collect. (In fact, free, if you already get the newspaper and magazines.) Begin clipping right away. You may wish to limit your collecting to name bears like Snuggle, Pooh, or Eckerd's Humfrey; collect just those bears that especially appeal to you, or include them all. Don't forget to save all advertising premium offers, even if you don't send for the item. (Original premium offers and ads featuring the advertising bear increase the bear's value.) Organize ads alphabetically in clear open-topped sheet protector pages in a notebook. You'll soon have a wealth of knowledge and reference material to make your advertising bear collecting more fun. Below is a sampling of what you can expect to find.

Of course, another facet of bears on paper is storybooks about bears, but that would require an entire other book to cover.

A "take away" Kids Meal paper bag from A&W Root Beer featuring the Great Root Bear, of course.

Baseball, Coneys, and root beer go together. In this A&W Restaurant ad, from a Washington State newspaper, the Great Root Bear is ready with his bat in hand.

Klondike Bear tempts us with his delightfully delicious ice cream wares.

It's the Klondike Bear again, letting us know about some cool new ice cream treats.

Pooh's giving us a "hunny" of a deal, with the purchase of Oral B toothbrushes, $5 off on his video and free floss.

Although the Campbell Kids are their usual "spokes folks," Campbell's Soup brought Pooh into the advertising fold by offering a free Pooh CD-ROM Sampler.

This ad offers "Winnie the Pooh the whole year through." It's a calendar with changeable pictures.

Telegrams may no longer be "state of the art" but "Pooh Grams" are. A Pooh bear, dressed for the occasion (birthday, graduation, I Love You, Get Well, It's a new baby girl/boy, or holidays) comes along with a book and card.

"Pooh's Letters from the Hundred Acre Wood" are detailed in this ad decorated with all the Pooh friends.

"Pooh's Little Baby Gift" book and rattle set are surrounded by the inhabitants of the 100 Acre Wood.

Snuggle asks us to "Cuddle up." Who could resist?

Snuggle says "If you love soft things, then you'll love me." My Mom used to love his TV commercials.

Snuggle invites us to "Get your clothes as soft as me!" A group of Snuggle ads framed and hung in a baby's room would be a cozy touch.

"can't bear to be without it" is a perfect statement for a spokesbear.

It's easy to see why Snuggle remains an active ad campaigner.

Here Healthtex clothing teams up with Snuggle Bear and the fabric softener to sell both products.

Here's another team-up for Snuggle. This time with Little League.

This Snuggle ad would frame nicely

*S*nuggle® can't care for all the world's colors. But it sure can help when it comes to clothes.

Can a fabric softener really change the way you see color? It can if it's Ultra Snuggle with Color Protector. It's brilliant the way this new Snuggle helps care for your colors. Its color-active ingredient actually helps keep some colors from redepositing on certain clothes. Feel the Snuggle softness, see the Snuggle difference for yourself. Available in refill, too.

Another good ad possibility for framing.

Sue Bee Cookies
'll Wanna Eat by the Paw

The Sue Bee Honey Bear is as friendly on paper as he is on television.

Sue Bee Bear is really a honey.

Another reason for saving Sue Bee Honey ads is for their recipes.

In late 1995 Post's Golden Crisp cereal enclosed Sugar Bear ornaments in specially-marked boxes of the product. Except for this ad I might not have known about them.

In 1996 Teleflora and Hershey's pooled their products for this delightful Valentine's Day gift. It was called the Hershey's Kisses Bear Bouquet.

Teleflora and Coca-Cola joined forces in 1996 for this dual gift suggestion.

Another 1996 Teleflora-Coca Cola joint effort. Note "Coca Cola" on the bear's red chest heart.

This polar sled bear is making tracks for bargains in the frozen food aisle.

Who wouldn't believe this little bear when he says it's "Real Soft?"

A back cover sales piece for a combination book/record from Peter Pan © in 1981.

A newspaper ad for a local restaurant. The outdoor sign would be a good possibility for your Bear Signs photo book too.

6. Product Packaging

Bears as collectibles have mushroomed into a booming business over the past ten to fifteen years. Manufacturers of numerous products have hopped on the gravy train to take advantage of this interest.

Almost daily, new products with bears on their packaging find their way to grocer's shelves. Many are introduced by accompanying magazine or newspaper glossy-insert ad campaigns. Some items you may be familiar with. Others may surprise you.

Collecting the packaging can turn your grocery shopping expeditions into enjoyable bear hunts. Most of the items are inexpensive. You can eat or use the package contents and build quite a collection of the labels, wrappings, or boxes (which can be stored easily when flattened.) Some whimsical ones you may want to frame.

Foods

Valentine's Day is more fun when a cute bear offers you bear hugs and gumballs, 1998.

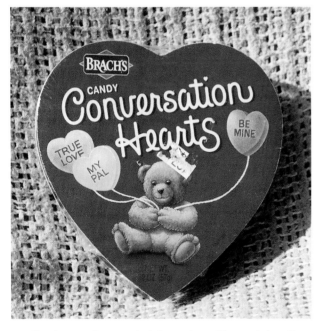

Brachs put a bear on their heart shaped box to help sell their "Conversation Hearts," 1998.

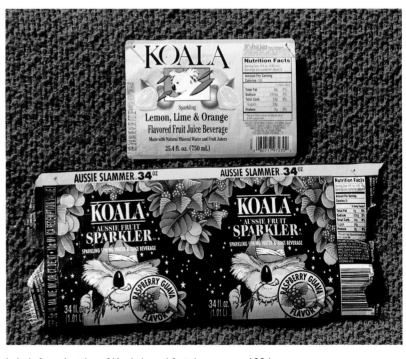

Labels from bottles of Koala brand fruit beverages, 1994.

Cans of Koala brand "Aussie Fruit Sparkler" beverages. Orange-Mango, Kiwi Lime-Grapefruit, and Raspberry-Guava flavors, 1994.

For Valentine's Day 1998, Hershey's packaged red and silver foil wrapped Kisses in a clear plastic tube topped by a plastic bear. Unfortunately only the tube says "Hershey's Kisses."

This little 3" plastic bear holds a heart shaped sign that says "Be Mine." She's the topper of the Hershey's Kisses tube. However, since she has no markings of any kind, future finders will have no clues to her advertising bear beginnings, 1998.

Palmer Candies also display bears on their candy packages, 1994.

Powers Candy knows that "bear power" helps move their multiple variety of candies - Jelly Bird Eggs, London Licorice, salt water taffy, coconut dips, sweet/sour gummi bears, robin eggs, chocolate bunny, and jelly rabbits. The packages shown are from March 1994. Note the bears appear near the company name.

Super Golden Crisp cereal box. Standing on top is the little fabric "Sugar Bear" who came in it. *Courtesy of Oneida Callaway, photo by Mary Ann Callaway Dennis.*

Back of the Post's Golden Crisp cereal box offering a Sugar Bear watch, sports bottle, and T-shirt.

Post's Golden Crisp cereal with Sugar Bear on the front.

Ralston's Frosted Flakes must be good if the lip-licking bear is to be believed, 1998.

Kountry Fresh Frosted Flakes seem to appeal to this polar bear, 1998.

I bet I'm not the only one that bought Kountry Fresh's Honey & Nut Toasted Oats cereal because of the darling climbing koala on the front, 1998.

Another koala bear tastes Publix Fruity Crisp Rice, 1998.

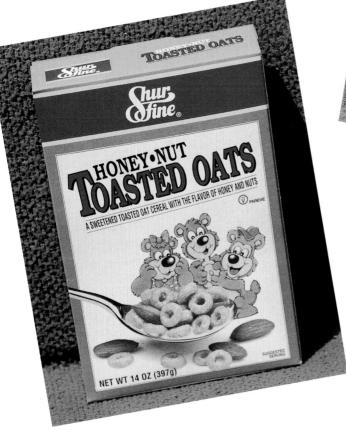

Shur-Fine's Honey Nut Toasted Oats box suggests it's for your whole family (of bears), 1998.

Maybe this panda has eaten Shur-Fine's Crispy Rice before, because he's already licking his lips, 1998.

The bear on Publix Cocoa Crisp Rice cereal looks like he might be related to Baloo, 1998.

Everyone knows bears love honey. This bear is adding it to Hyde Park's Honey Nut Toasted Oats, 1998.

Always Save ® Frosted Toasted Oats box front implies their cereal will give you the "oomph" to go skiing like this bear, 1997.

Here's Mama Bear on the Malt-O-Meal Golden Sugar Puffs package, 1994.

Tem-T Sugar Cones are tempting this bear, 1994.

This package gives you two bears for the price of one - The Klondike Bear and the Choco Taco Bear, 1998. (Never mind a camel, I'd walk a mile for one of these.)

Eskimo Pie is bringing back original 1949 ad campaign artwork on their packaging, beach towels, Tee shirts, and posters. These bears are really cute,1998.

Kraft's Cool Whip non-dairy whipped topping used this decorator container with bears at the beach for their product several years ago. The only marking on the bottom is "Top rack dishwasher safe."

Even generic products use bears on their packaging. This little white bear iceman is a cutie.

Celestial Seasonings Herb Teas feature bears on the packaging of several of their tea flavors. Here are Lemon Berry Zinger, Harvest Chamomile, and Lemon Lime Splash Iced Delight, 1997.

Two more plastic honey containers. The bear with the tie says "Harkey's Honey, San Saba, Texas." The jug is for "Joe's Honey, Loxahatchee, Florida," 1994.

Honey and bears are naturals together. Sue Bee® Honey, Sioux Honey Association, uses a brown bear in their magazine and TV commercials, and this little plastic bear container, 1994.

Ross's infant formula Similac and Isomil assign their bears different colored bows - green, yellow, blue, and red - for instant recognition of each formula type, 1994.

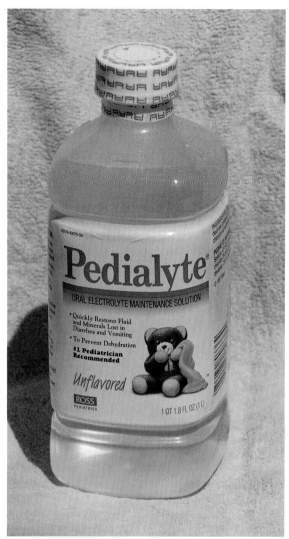

Ross Labs Pedialyte also pictures a little bear with his blue bow and blue blanket, 1998. *Photo courtesy of Peggy Monahan.*

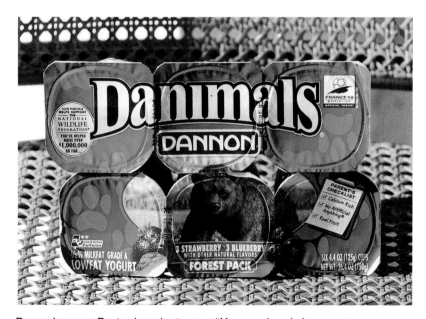

Dannon's yogurt Danimals packaging says "Your purchase helps support the National Wildlife Federation. You've helped raise over $1,000,000 so far." A brown bear appears on the "Forest Pack" strawberry-blueberry flavor, 1998.

Non-Foods

Crayola offered this holiday pack with 24 crayons and a bonus of a bear tree ornament in 1997.

This little Crayola Scribble Pad bear can help me color anytime, 1997.

Prang is a new contender in the crayon market, making crayons from soy beans and using Pandas on their packaging.

Kodiak Snuff features a "manly" brown bear on the can tops, and offered a T-shirt with a picture of their spokesbear.

This box of Kleenex ® facial tissue has a bear on all four sides. On one side Mother Goose is pointing out to him the stars in the night sky. Another side shows him catching a shooting star. Another side is a bear rafting trip down the river in candyland.

Angel Soft bathroom tissue uses both a bear and baby to entice us into buying their product.

Although Pampers diapers pictures a baby on their packages, a helpful little bear gives extra information like "New, Gentle Touch Liner, with Aloe."

Kash n' Karry scented baby wipes also "star" a golden bear on their plastic dispenser boxes.

Chubs stackables baby wipes have been using a little golden bear on their packaging for a number of years.

7. Bearly Edibles
Bear-shaped foods

Bear-shaped foods are not a new concept. Gummi bears have been around for decades. Smokey Bear fans may be surprised to hear that in 1985 Timberline Quality Confections made Smokey Bear Gummi Candies. Cookbooks, such as *Holiday Celebrations* (© 1997 Publications International Ltd, Lincolnwood, IL.) and old copies of women's magazines reveal ideas and instructions for homemade foods shaped like bears.Browse through your cookbook library; you may "bag a bear."

Recipes, instructions, and photos of bear-shaped breads, cakes, and cookies can be found in the following cookbooks and magazine articles:
- "Christmas Bear (bread) Filled with Clam Dip," pg 66-67, *Bridgford® Home Baking Collection*, © 1996, Bridgford Foods Corporation, PO Box 773, Anaheim, CA 92803.
- "Honey Bear Breads," pp 92-93, *Pillsbury Family Christmas Cookbook*, © 1991 Pillsbury Co., Doubleday, New York.
- "Teddy Bear (Face) Cake," (made using one round cake pan) from "Funny Food," pg 28, *Military Lifestyle magazine*, April 1993. Recipe from *Better Homes and Gardens Kids' Party Cook Book*, © 1985 by Meredith Corp., Des Moines, Iowa.
- "Teddy Bear Cake," (supine bear made using 1 square and 1 round cake pan) an advertisement for Baker's Chocolate and Baker's Angel Flake Coconut, pg 113, December 1968, *Better Homes and Gardens* magazine. The ad appeared in several women's magazines in 1964-65.

- "Ted Bear Cake," from "Animal Fair," pgs 77, 102-103, March 1975, *Better Homes and Gardens* magazine. The upright seated bear cake is made using a 2 lb metal coffee can (body), a 3 lb metal shortening can (head), four 10-oz metal soup cans (legs & ears), muffin pan (1 cupcake for nose).
- "Apricot Panda"Cake, (supine panda made using 1 round and 1 square cake pan), from "Perfect Party Cakes," pg 69, July 1974, *Woman's Life* magazine.
- A 3-D, seated Panda Bear Cake pan was advertised by Wilton and pictured in their *Wilton's Food and Cake Decorating Year Books* of the 1970s.
- "Teddy Bear Cookie," (large bear sugar cookie standing upright in a sugar cookie gift box), pgs 96-97, 105-106, December 1981, *Ladies Home Journal* magazine.

Making any of these "bearly edible" bruins will guarantee you bear hugs from the delighted recipients.

Due to the onset of "bear mania," scores of manufacturers are producing foods in the shape of bears. Bear lovers of all ages enjoy them. Labels or boxes usually feature bears with a storybook look, sometimes whole families of them. Here are a few examples. (Remember to save the labels and flattened boxes for your Bear Pals on Paper advertising/packagaing notebook.)

On the left a 15 oz can of Franco-American SpaghettiOs ™ TeddyOs ™, "Fun bear shapes, Same great taste" Pasta in tomato & cheese sauce. Label shows gold bear eating. On the right is a 10.5 oz can of Campbell's condensed soup "Teddy Bear pasta shapes with chicken broth." Label depicts bear on a skateboard. (See Bears A-Z for plush Campbell's Teddy Bear Soup bear.)

Kraft Macaroni & Cheese Dinner, Teddy Bears, 5.5 oz box, three bear children on box front.

Mueller's Super Shapes, Teddy Bears, 11.5 oz box. Box front depicts a bear boy standing on another bear boy's shoulders to swat a hornet's nest out of a tree and a girl bear is sitting down laughing. The back of the box has puzzles and games.

Kraft Macaroni & Cheese Dinner, Teddy Bears, 5.5 oz box, bear at the beach on box front. 1994

Pillsbury Teddy Bear and Dinosaur refrigerated dough for cookies. Although they are brown and white, they are sugar cookies with no chocolate.

Sunshine Grahamy Bears, "honey graham crackers in little bear shapes," 12 oz box. Bear on box front is delving into a honey pot with a spoon.

The Original Black Forest Gummy Bears "won't stick to your teeth." 5.2 oz package with blue skies, tall green trees, and a brown bear with a beige oval on his tummy. These Gummy Bears were distributed by the Foreign Candy Company in Hull, Iowa, incorporated in 1982.

"Haribo Gold-Bear Minis, The Original Gummi-Bears, 10 oz., Product of West Germany" Inside the bag are individual sealed packages that hold 8 gummi-bears.

The Berry Bears, chewy fruit snacks, fruit punch, made with real fruit, came packed inside in 6 individual pouches. By Fruit Corners ®, General Mills, Inc. On the box back is a story about having a birthday party for their horse friend "Flight." The box front shows sister, brother, and diapered baby enjoying their fruit snacks. Made by Betty Crocker.

A 1.5 oz package of "Olde World Gummy Snacks, made with real fruit juices, Gummy Bears." The smiling bear on the package front is red and is dressed in only his white gloves. Brock Candy Co., Chattanooga, Tennessee.

Nabisco Chocolate Teddy Grahams, graham snacks, 10 oz box. Tiny bite-size chocolate bears appear to spill out of the box front. (A magazine article titled "Guilt Free Sweets" listed chocolate Teddy Grahams as 60 calories and 2 grams of fat for 11 of the little bears.)

Back of the Teddy Grahams box shows the "four delicious flavors" the little bears come in: honey, vanilla, chocolate, and cinnamon. "Mother's Little Helper, Teddy Grahams, The delicious way to bring wholesome snacks out of hibernation."

8. Other Beary Good Stuff

Some collectors stick strictly to plush bears and don't "get into" other items featuring bears. For those of you who can't resist bears in any form, the following samples may provide some ideas. You may even decide to start an adjunct bear collection of advertising mugs, tins, or conversation pieces. Or postcards, postmarks, travel folders from towns or places with bears (e.g. Big Bear Lake, Bear Mountain) in their names. Miniature sports team pennants or tourist flags featuring bears, for your little bruins to hold, is an idea for an accessory collection that requires only a modicum of space.

A&W Great Root Bear Flyer frisbee. $4-$8

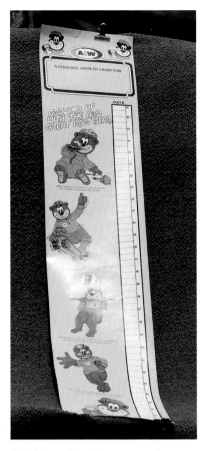

A&W Great Root Bear personal growth chart, 35" long, "© 1992 A&W Restaurants." $5-$10.

A plastic rain poncho from Baylor University shows their growling bear mascot. Baylor was a charter member of the Southwest Conference in 1914. BU has two "live" bears that appear at football games and campus events. (Wonder if that means costumed or trained??) Poncho $4-$10. (Price may be subject to how hard it's raining at the time.)

A jacket or shoulder patch "Member," smiling bear holding a long pole, "Bear, 500 Club." It's likely that it refers to engines or auto racing. Patch $2-$6.

Atop the Celestial Seasonings cardboard tea box sits a tin replica of the box. Tin "© 1982" To the right is a ceramic lidded cup with a full color Sleepy Time bear sleeping in his chair. *Courtesy of Jean Laughery.*

COKE® BEAR SLIPPERS

A fuzzy fun pair of Coca-Cola® polar bear slippers with attached red scarves and Coke® tush buttons. They were available through a gift catalog in 1995 for $24.95.

5" doll size pink plastic divided plate with two bears and "Color Magic Cooking." Incised on the plate bottom "Thailand." $1-$2.

Hamm's Bear lookalike black mesh billed cap. Tag: "Designer Award Cap, Taiwan ROC." Original price $13.50.

Hershey's wooden Christmas tree ornament. The little man is trying to extricate a Hershey's bar from the Hershey's mailbox. (Or maybe he's looking for a Hershey's Kiss bear.) The paper tag is also a replica of a Hershey's bar. $3-$10.

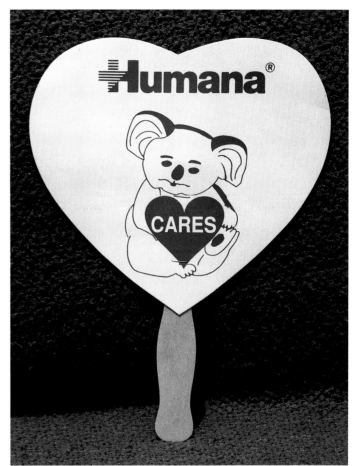

A light weight cardboard hand fan with a wooden handle lets you know "Humana ® Cares." $1-$3.

This 7" tall rubbery bear is a definite "what's-it." There's a 3" hole in his bottom, and a 2" hole in the bucket he's holding. A real icebreaker conversation piece. The only marking is "Made in China" incised on his pants. $1-$5.

Ceramic "perfume bottles." Bear in chef's cap says "Lucy Lee Perfume" Bear with headdress like an Elizabeth Taylor "Sapphires, Emeralds, Rubies" perfume bottle, says "Parfum Noelle." They are ceramic figurines, not perfume bottles, marked on the bottom: "Lucy & Me, © 1995 Lucy Rigg, Licensee Enesco, Made in China." The "Lucy & Me Collection" was introduced by Enesco in 1978, and now contains more than 1,000 figures. Artist Lucy Rigg began creating clay bears in 1969 while awaiting the birth of her daughter, Noelle. She first sold her hand-painted bears, which she called "Rigglets," at street fairs. There is a Lucy & Me collectors club. Bear "perfume bottles" $10-$35 each.

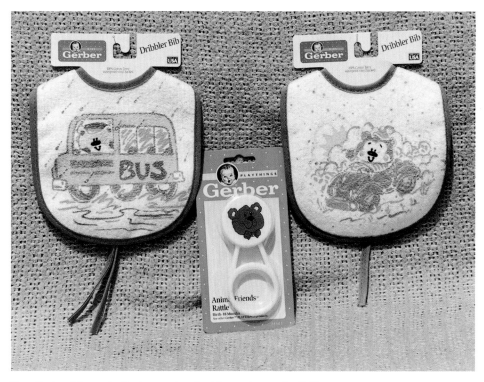

Two Gerber bibs and a Gerber rattle, all with cute bears. Available at grocery stores.

Mr. and Mrs. Bear go for a ride in a Johnson & Johnson truck (10" long x 6.5" wide.) Lift where it says "Johnson & Johnson." It tilts back like a dump truck exposing a storage space. No markings on the truck. Pegs on truck may have held passenger bears. $5-$8 as is.

Green cap with a message "Kids Need Hugs Not Drugs." Two outlined bears demonstrate hugging. $3-$7.

Bibliography

Bear-in-Mind catalog, Concord Massachusetts, Holidays 1993.

"Cooperstown's Ball Park Bears," p 86, *Teddy Bear and friends*, November/December 1994.

Cooperstown Bears catalog, Buffalo Grove, Illinois, 1998.

Douglas Cuddle Toys, *Company Classics catalogs*, Keene, New Hampshisre, 1995, 1998.

Hake, Ted, *Hake's Guide to Advertising Collectibles, 100 Years of Advertising from 100 Famous Companies*, Wallace-Homestead Book Company, Radnor, Pennsylvania, 1992.

Hockenberry, Dee, "Teddy Bear Price Guide," *Teddy Bear and friends*, January/February 1994, p. 37.

Johnson, Jo Ann, "Be on the Lookout! AppealingLogo Bears Have Proven to be Good Investments," *Teddy Bear and friends*, March /April, 1993, pp 98-100.

Kletzel, Dani, "How It All Begund." *Teddy Bear and friends*, July/August, 1994, p. 26.

Lamphier, Mary Jane, *Zany Characters of the Ad World, Collector's Identification & Value Guide*, Paducah, Kentucky, Collector Books, 1997.

Machir, Virginia Hearn, "Advertising Dolls of the Past, Lee Teddy Bear," *Doll World*, April, 1993, p. 43.

Machir, Virginia Hearn, "Advertising Dolls of the Past, Shoney Bear," *International Doll World*, 1990, p. 14.

Reed, Robert, "Checking Out the Candy Bar Connection," *Doll World*, February, 1996, pp.10-11.

Robinson, Jolee Ashman and Kay F. Sellers, *Advertising Dolls, Identification & Value Guide*, Paducah, Kentucky, Collector Books, 1980.

Rohland, Pamela, "The Cuddle Company," *Teddy Bear and friends*, January/February 1996, p. 76.

"Sleepy Travelodge Bear," *Doll Collector's Price Guide*, Summer 1997, p.42.

Index

PEZ® Collectibles Expanded 2nd Edition Richard Geary. After PEZ candy and dispensers were brought to the United States in 1952, over 250 different types of character heads were made. Licensed cartoon characters, movie and T.V. personalities, and original designs are divided chronologically into five series, usually indicated by patent dates on the dispensers. The premiums and store displays are shown too.

Size: 8 1/2" x 11"   Price Guide   112pp.
281 color photos
ISBN: 0-7643-0315-5   soft cover $19.95

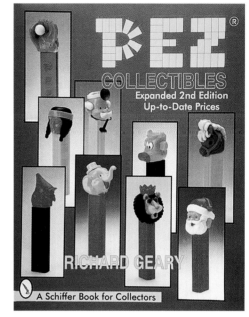

More Pez Expanded and Updated 2nd Edition Richard Geary. The popularity of PEZ collectibles has never been higher. In part this is due to Richard Geary's first book, PEZ Collectibles. Now he offers those sufferers from "PEZ-O-Mania," new discoveries and new PEZ products.

Illustrated with hundreds of full color photographs, this new book will contribute more knowledge to the field as well as an interest source for tracking a collection. An accurate price guide is included.

Size: 8 1/2" x 11"   150+ color photos   80 pp.
Price Guide
ISBN: 0-7643-0453-4   soft cover $14.95

The Collector's World of M&M's™: An Unauthorized Handbook and Price Guide Patsy Clevenger. Here's a sweet treat for collectors of colorful M&M's characters, starring regular and peanut. This is a serious collector's guide to those smiling candies with a tough exterior and a gooey heart. Author Patsy Clevenger, who describes the smiles of the M&M characters as "absolutely infectious," first provides readers with a concise review of the candy's history, including a timeline of M&M colors featured from 1940 through the 1990s. This comprehensive and enjoyable guide then leads the reader on a tasty tour of M&M collectibles, complete with values for the hundreds of items shown, which range from kitchen towels to tins, toppers, posters, and jewelry. Additional sections on advertising items and M&M packaging round out the book.

Size: 6"x 9"   298 color photos   128 pp.
Price Guide
ISBN: 0-7643-0406-2   soft cover $16.95

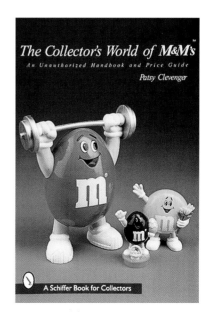

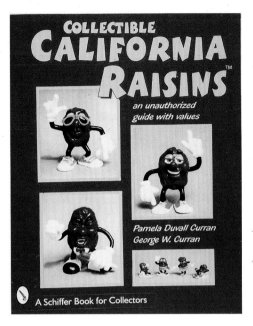

Collectible California Raisins™: An Unauthorized Guide with Values Pamela Duvall Curran, George W. Curran. California Raisins™ lovers no longer need rely on the grapevine for collector information. Here is the authoritative illustrated guide, complete with more than 500 color photographs of collectible items and their values. Born of popular advertising and projected to near stardom, the California Raisins™ held sway in the United States for nearly a decade. Begun as an advertising venture to promote a farm product, the Claymation® characters actually became their own industry. Starting with figural characters distributed in everything from fast food to fast car promotions, the California Raisins™ soon appeared on everything imaginable, from bed linens to wallets. They even starred in their own books, games, and, of course, videos. The entire range of Raisins collectibles is represented in this captivating and irresistible work.

Size: 8 1/2" x 11"	509 color photos	128 pp.
Price Guide		
ISBN: 0-7643-0433-X	soft cover	$19.95

Planters Peanut™ Collectibles, 1906-1961: A Handbook and Price Guide Jan Lindenberger. Over 100 years ago, in 1896, Amedeo Obici bought a peanut roaster from his uncle and went into business for himself. After ten years of selling peanuts to other shopkeepers from a small store and a horse-drawn cart, in 1906 Obici teamed up with Mario Peruzzi, to expand business, and form Planters Peanuts. The rest is history. Planters Peanuts is a part of American culture, and its innovative advertising has produced a large and vital group of Planters collectors. Color photographs documents the early years of Peanuts memorabilia, & with descriptions and prices, it is a great and much needed guide for collectors.

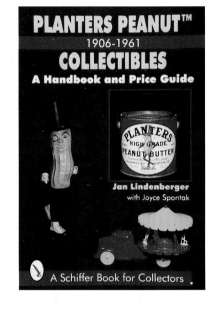

Size: 6" x 9"	over 450 color plates	160 pp.
Price Guide		
ISBN: 0-88740-792-7	soft cover	$19.95

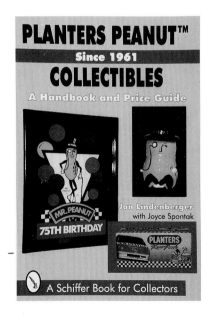

Planters Peanut™ Collectibles Since 1961: A Handbook and Price Guide Jan Lindenberger. In 1961 Planters Nut and Chocolate Company was acquired by Standard Brands, but continued to produce a great number of advertising premiums. These "newer" items of Planters memorabilia are presented here, including those surrounding its involvement in auto racing and golf tournaments and the promotional items surrounding Mr Peanut's 75th birthday in 1991. This is a companion volume to Planters Peanut Collectibles: 1906-1961.

Size: 6" x 9"	500 color photos	160 pp.
Price Guide		
ISBN: 0-88740-793-5	soft cover	$19.95